Robust Multivariable Flight Control

Richard J. Adams, James M. Buffington,
Andrew G. Sparks and Siva S. Banda

Robust Multivariable
Flight Control

With 98 Figures

Springer-Verlag
London Berlin Heidelberg New York
Paris Tokyo Hong Kong
Barcelona Budapest

Richard J. Adams, BS, MS
James M. Buffington, BS, MS
Andrew G. Sparks, BS, MS
Siva S. Banda, BS, MS, PhD

WL/FIGC Bldg. 146
2210 Eighth Street, Suite 21
Wright-Patterson AFB
OH 45433–7531
USA

ISBN-13: 978-1-4471-2113-8   e-ISBN-13: 978-1-4471-2111-4
DOI: 10.1007/978-1-4471-2111-4

British Library Cataloguing in Publication Data
Adams, Richard J.
  Robust Multivariable Flight Control. –
  (Advances in Industrial Control)
  I. Title II. Series
  629.8312

Library of Congress Cataloging-in-Publication Data
Robust multivariable flight control / Richard J. Adams . . . [et al.].
      p.   cm. — (Advances in industrial control)
  Includes bibliographical references and index.
                    (Berlin : acid-free paper). — ISBN
0–387–19906–3 (New York : acid-free paper) :
  1. Flight Control.   2. Nonlinear control theory.   I. Adams,
Richard J., 1967–   .   II. Series.
TL589.4.R63 1994   94–3683
629. 135—dc20   CIP

© Springer-Verlag London Limited 1994
Softcover reprint of the hardcover 1st edition 1994

The publisher makes no representation, express or implied, with regard to the accuracy
of the information contained in this book and cannot accept any legal responsibility or
liability for any errors or omissions that may be made.

Typesetting: Camera ready by authors

69/3830–543210 Printed on acid-free paper

SERIES EDITORS' FOREWORD

The series *Advances in Industrial Control* aims to report and encourage technology transfer in control engineering. The rapid development of control technology impacts all areas of the control discipline. New theory, new controllers, actuators, sensors, new industrial processes, computing methods, new applications, new philosophies, . . . , new challenges. Much of this development work resides in industrial reports, feasibility study papers and the reports of advanced collaborative projects. The series offers an opportunity for researchers to present an extended exposition of such new work in all aspects of industrial control for wider and rapid dissemination.

Robust control theory has now reached the stage where practical implementation as prototype controllers are appearing. The high performance demands of aerospace flight control are a natural application for H_∞ and structured singular value robust control. Dr. Banda and his colleagues at Wright Patterson Airforce Base, Ohio, U.S.A. had long recognised the aerospace applications potential in the new theory. This volume reports the way in which the robust methodology has to be modified to suit the special features of the system for which controllers are being designed.

Two design case studies are presented for the manual flight control of lateral/directional axes of the VISTA-F-16 test vehicle and an F-18 trust vectoring system. The interplay between theory and the physical features of the systems is self-evident and forms a significant feature of the work presented. Nonlinear system simulations demonstrate the satisfactory performance achieved. Altogether an instructive and invaluable addition to the Advances in Industrial Control Series.

M. J. Grimble and M. A. Johnson
Industrial Control Centre
University of Strathclyde
Scotland, U.K.

Acknowledgements

The work presented in this monograph was done under work unit number 2304N201. The authors would like to thank the Air Force Office of Scientific Research for their sponsorship of this effort. This work was performed by the Control Analysis Section, Control Dynamics Branch, Flight Control Division, Flight Dynamics Directorate, Wright Laboratory, Wright-Patterson Air Force Base, Ohio, The United States of America.

CONTENTS

EDITORIAL BOARD

ACRONYMS, ABBREVIATIONS, AND SYMBOLS

Acronyms AND Abbreviations

cg	Center of Gravity
CAP	Control Anticipation Parameter
LQ	Linear Quadratic
V/STOL	Vertical/Short Take-off and Landing
LOES	Low Order Equivalent System
HOS	High Order System
VISTA	Variable Stability In-Flight Simulator Test Aircraft

Symbols

A	System Matrix
B	Control Effectiveness Matrix
B^*	Generalized Control Effectiveness Matrix
C	Output Matrix
CS	Control Selector
D	Direct Feed-through Matrix
g	Gravity
h	altitude
I	Identity Matrix
I_x	Inertia About Aircraft X-body axis
I_{xz}	Cross Product of Inertia
I_y	Inertia About Aircraft Y-body axis
j	$\sqrt{-1}$
L	Roll Moment
L_β	Derivative of Roll Moment wrt Sideslip Angle
L_δ	Derivative of Roll Moment wrt Control Deflection
L_p	Derivative of Roll Moment wrt Roll Rate
L_r	Derivative of Roll Moment wrt Yaw Rate
M	Pitch Moment
m	mass
M_α	Derivative of Pitch Moment wrt Angle of Attack
M_δ	Derivative of Pitch Moment wrt Control Deflection
M_q	Derivative of Pitch Moment wrt Pitch Rate
M_u	Derivative of Pitch Moment wrt Forward Velocity
N	Yaw Moment

N_β	Derivative of Yaw Moment wrt Sideslip Angle
N_δ	Derivative of Yaw Moment wrt Control Deflection
N_p	Derivative of Yaw Moment wrt Roll Rate
N_r	Derivative of Yaw Moment wrt Yaw Rate
n_z'	Normal Accel at Instantanious Pitch Center of Rotation
P	Total Roll Rate
p	Perturbational Roll Rate
$\dot{p}_c$	Generalized Roll Acceleration Command
Q	Total Pitch Rate
q	Perturbational Pitch Rate
$\dot{q}_c$	Generalized Pitch Acceleration Command
$\bar{q}$	Dynamic Pressure
R	Total Yaw Rate
r	Perturbational Yaw Rate
$\dot{r}_c$	Generalized Yaw Acceleration Command
Re	Real Part
s	Laplace Operator
T	Control Selector Transformation
T_r	Roll Mode Time Constant
T_s	Spiral Mode Time Constant
U	Total Velocity Along X-body axis
u	Perturbational Velocity Along X-body axis
V	Total Velocity Along Y-body axis
V_T	Total Velocity Along the Flight Path
v	Perturbational Velocity Along Y-body axis
v^d	Desired Eigenvector
W	Total Velocity Along Z-body axis
w	Perturbational Velocity Along Z-body axis
X	Total Force Along X-body axis
x	Perturbational Force Along X-body axis
x	State Vector
X_α	Derivative of Force Along X-body axis wrt Angle of Attack
X_δ	Derivative of Force Along X-axis wrt Control Deflection
X_u	Derivative of Force Along X-body axis wrt Forward Vel
Y	Total Force Along Y-body axis
y	Perturbational Force Along Y-body axis
Y_β	Derivative of Force Along Y-body axis wrt Sideslip Angle
Y_δ	Derivative of Force Along Y-axis wrt Control Deflection
Z	Total Force Along Z-body axis
Z_α	Derivative of Force Along Z-body axis wrt Angle of Attack
Z_δ	Derivative of Force Along Z-axis wrt Control Deflection
Z_q	Derivative of Force Along Z-body axis wrt Pitch Rate
Z_u	Derivative of Force Along Z-body axis wrt Forward Velocity
α	Angle of Attack
β	Angle of Sideslip
δ	Control Deflection
δ^*	Generalized Control Deflection

δ_c^*	Generalized Control Deflection Command
δ_A	Aileron Deflection
δ_{DF}	Differentail Flaperon Deflection
δ_{DT}	Differentail Horizontal Tail Deflection
δ_E	Symmetric Horizontal Tail Deflection
δ_{lat}	Lateral Stick Deflection
δ_p	Pitch Stick Deflection
δ_{ped}	Rudder Pedal Deflection
δ_{PTV}	Pitch Thrust Vectoring Nozzle Deflection
δ_R	Rudder Deflection
δ_{RTV}	Differential Pitch Thrust Vectoring Nozzle Deflection
δ_{YTV}	Yaw Thrust Vectoring Nozzle Deflection
Δ_m	Relative Error
Φ	Total Roll Euler Angle
ϕ	Perturbational Roll Euler Angle
λ^d	Desired Eigenvalue
μ	Structured Singular Value
$\dot{\mu}$	Stability Axis Roll Rate
ν	Desired Dynamics for Dynamic Inversion
Θ	Total Pitch Euler Angle
θ	Perturbational Pitch Euler Angle
ρ	Spectral Radius
$\bar{\sigma}$	Maximum Singular Value
τ_θ	Equivalent Pitch Time Delay
ω	Frequency
ω_p	Phugoid Natural Frequency
ω_{sp}	Short Period Natural Frequency
Ψ	Total Yaw Euler Angle
ψ	Perturbational Yaw Euler Angle
ζ	Damping
ζ_p	Phugoid Damping
ζ_{sp}	Short Period Damping
$\| A \|_p$	p-norm of A
A^T	Transpose of A
A^{-1}	Inverse of A
$A^\#$	Generalized Inverse of A
$\Re$	Set of Real Numbers
C	Set of Complex Numbers

Subscripts/Superscripts

long	Longitudinal
lat/dir	Lateral/Directional
aero	Aerodynamic
tvec	Thrust Vectoring
∞	Infinity

CHAPTER 1

INTRODUCTION

Manual flight control system design for fighter aircraft continues to be one of the most demanding problems in the world of automatic control. It is the job of the manual flight control system to provide a satisfactory dynamic response to pilot inputs. The problem is inherently multivariable, that is, a controller must drive multiple effectors based on information from multiple sensor and command inputs. Modern fighter aircraft dynamics generally have highly coupled dynamics that are both uncertain and nonlinear.

In the case of such multivariable problems, the only efficient means of obtaining a solution is to use multivariable control design techniques. Classical control techniques that close a single feedback loop at a time are inefficient for designing control systems for plants with multiple inputs and outputs and strong coupling between the loops. Further, classical techniques do not address robustness in multivariable systems adequately, as they only address stability margins in single loops and neglect simultaneous perturbations in several loops. Multivariable control techniques can provide an efficient means of finding control laws for complex systems and make it straightforward to incorporate specific performance and robustness requirements.

The magnitude of the manual flight control problem is driven by the nonlinear and uncertain nature of aircraft dynamics. Linear models of these systems are only valid for small regions about trim conditions. The conventional solution to this problem is to perform point designs for a large set of trim conditions and then construct a gain schedule by interpolating gains with respect to flight condition. This ad hoc procedure is time consuming and expensive, but is well accepted and has yielded satisfactory results for dozens of aircraft.

One advantage of classical controllers is that they are straightforward to schedule with flight condition because there is a clear meaning for each one of the gains or dynamic filter elements. The use of multivariable control in full envelope flight control law design is difficult because the resultant controllers are usually dynamic compensators in state space form and often are of high order. Such control laws are extremely difficult to schedule with varying flight condition, as the elements in the state space description of the controller do not vary smoothly with flight condition, and their relation to any particular parameter is not clear.

It is tempting to represent the variation in aircraft dynamics with flight condition as a model uncertainty and design a single, fixed control law for a nominal aircraft model. This would eliminate the need to schedule the controller with flight condition. However, making the control law robust to all changes in the dynamics due to changes in flight condition would mean sacrificing performance. No single, fixed control law will give maximum performance across a wide flight envelope. In addition, although the dynamics at a particular flight condition are usually uncertain to some degree, there is usually a reasonable amount of information about how the flight dynamics change as the operating condition of the vehicle changes. Some alternative to representing the plant parameter variation as model uncertainty is required.

An obvious solution to this problem is to design constant gain multivariable controllers whose elements can be scheduled with flight condition. This is a useful alternative in some cases, although the number of techniques available for this are somewhat limited, and in some cases these have restrictive assumptions on the design model such as requiring state feedback. Furthermore, many of the most powerful robust control design approaches, which account for modeling uncertainty using frequency dependent bounds and guarantee robustness via the small gain theorem, produce dynamic controllers. While it is becoming more straightforward to obtain multivariable controllers for linear time invariant plants that meet specific performance and robustness requirements, using multivariable control as a tool in the overall design of a nonlinear system by individual point designs is still an issue.

Recently more attention has been given to applying a more mathematical basis to the issue of gain scheduling flight control systems. In [1.1], the author transforms plant dynamics into a quasi-linear parameter varying form in which dynamics depend on exogenous variables that are unknown but can be measured. This transformation allows explicit gain scheduling relationships to be derived which preserve stability in the presence of time varying plant parameters. In [1.2], the author describes a systematic approach for automating gain schedule calculations. The approach guarantees both stability and performance using structured singular value theory.

There are a few examples of attempts to gain schedule multivariable controllers in the literature. In [1.3], the authors designed H_∞ control laws at four widely spaced operating points for a pitch axis autopilot for a highly maneuverable missile. The plant dynamics are defined as an explicit function of some parameters such as angle of attack and dynamic pressure. The linear controllers at the different equilibrium points are interpolated depending upon the operating condition. Unfortunately, describing the aircraft dynamics as an explicit function of all its parameters is not a straightforward approach. In [1.4], the authors design H_∞ controllers for a V/STOL aircraft at discrete points and then switch between them based upon operating condition. Finally, in [1.5] and [1.6], the authors design flight control laws for a test aircraft using integral LQ regulators. The controllers found were constant gains plus integrators, and were scheduled as a function of dynamic pressure.

An alternative to gain scheduling is to use control design methods which directly consider the nonlinear nature of the problem. Adaptive control has received much attention recently and shows promise. Nonlinear dynamic inversion has been applied successfully to a number of flight control problems [1.7, 1.8, 1.9].

In this report, the problem of multivariable gain scheduling is addressed by using an inner/outer loop control law structure that separates the issues of scheduling the controller and designing control laws to meet specific performance and robustness requirements. The inner loop consists of a control selector that transforms generalized control commands into actual control effector commands and feedback compensation that minimizes the relative error between the inner loop dynamics at different flight conditions. Both the control selector and the inner loop feedback are functions of flight condition. The idea is to use the inner loop to make changes in the aircraft dynamics transparent to the outer loop. The outer loop consists of a fixed dynamic compensator to meet the performance and robustness requirements of the system.

Two design examples are presented to illustrate the effectiveness of the inner/outer loop approach. Manual flight control systems are designed for the lateral/directional axes of the VISTA F-16 test vehicle and for the longitudinal and lateral/directional axes of an F-18 with thrust vectoring. These two vehicles bring out a number of interesting flight control problems including effector limiting, thrust vectoring, and redundant controls. Three different inner loop design approaches are used to explore equalization methods. Dynamic inversion is used for the F-16 inner loop design, a reduced order H_∞ method is used for the longitudinal F-18 inner loop design, and eigenstructure assignment is used for the lateral/directional F-18 inner loop design. Flying qualities are built into each design using an implicit model following μ-synthesis outer loop formulation. An ideal model of the aircraft response which represents the desired flying qualities is included in the synthesis model. The infinity norm of the transfer function from the command input and the frequency weighted error between the ideal model response and the actual model response is minimized. Performance and robustness analysis is performed for wide ranges of operating conditions for each of the designs. Nonlinear simulation results are shown to demonstrate that the inner/outer loop design approach yields controllers that perform well in a highly dynamic and nonlinear environment.

Chapter 2 of this document presents the basic technical background behind the manual flight control problem and the theory used in the applications. It serves as an overview of the tools that are available to the designer of a multivariable flight control system. Chapter 3 introduces the inner/outer loop robust control design methodology. This methodology represents a framework for combining the wide range of available tools into a useful approach for deriving full envelope designs. Chapter 4 is a detailed application of this methodology to the design of a lateral/directional manual flight control system for the

VISTA F-16 aircraft. Chapter 5 is a detailed application of this methodology to a thrust vectoring F-18 manual flight control problem. These two chapters bring the design and analysis tools of Chapter 2 together with the methodology of Chapter 3 and demonstrate the utility of this approach for realistic fighter aircraft. Chapter 6 is a conclusion which sums up the results and recommends future research directions.

1.1   References

[1.1] J. S. Shamma and M. Athans, "Gain Scheduling: Potential Hazards and Possible Remedies," *IEEE Control Systems Magazine*, vol. 12, pp. 101-107, June 1992.

[1.2] R. Eberhardt and K. A. Wise, "Automated Gain Schedules for Missile Autopilots Using Robustness Theory," *Proc. 1992 IEEE Conf. on Control   Applications*, Dayton, OH, Sep, 1992.

[1.3] Robert T. Reichert, Robert A. Nichols, and Wilson J. Rugh, "Gain Scheduling for H_∞ Controllers: A Flight Control Example," *Proc. 1992 American Control Conf.*, Chicago IL, Jun. 1992.

[1.4] Richard Hyde and Keith Glover, "VSTOL Aircraft Flight Control Design Using H_∞ Controllers and a Switching Strategy," *Proc. 1990 24th IEEE Conf. Decision Contr.*, Honolulu, HI, Dec., 1990.

[1.5] Andrew Sparks, Richard Adams, and Siva Banda, "Control Law Development for the Lateral Axis of a Fighter Aircraft," *Proc. 1992 AIAA Guidance, Navigation, and Control Conf.*, Hilton Head, SC, August, 1992.

[1.6] Richard Adams, Andrew Sparks, and Siva Banda, "Full Envelope Multivariable Control Law Synthesis for a High Performance Test Aircraft," *Proc. 1992 IEEE Conf. on Control Applications*, Dayton, OH, Sep, 1992. Also, *Journal of Guidance, Control, and Dynamics*, vol 16 no 5, pp. 948-955, Sep-Oct 93.

[1.7] D. Enns, "Robustness of Dynamic Inversion vs. μ-Synthesis: Lateral-Directional Flight Control Example," *Proc. 1990 AIAA Guidance, Navigation, and Control Conf.*, Portland, OR, Aug. 1990.

[1.8] S. H. Lane and R.F. Stengel, "Flight Control Design Using Non-linear Inverse Dynamics," *Automatica*, vol. 24, pp. 471-483, 1988.

[1.9] C. Huang, et al., "Analysis and Simulation of a Nonlinear Control Strategy for High Angle of Attack Maneuvers," *Proc. 1990 AIAA Guidance, Navigation, and Control Conf.*, Portland, OR, Aug. 1990.

CHAPTER 2

TECHNICAL PRELIMINARIES

This chapter provides background information and describes the tools that are available to the designer of multivariable flight control laws. The fundamentals of aircraft dynamics are reviewed, the basics of flying qualities requirements are discussed, uncertainty models and analysis techniques are presented, and compensator design and order reduction methods are addressed. Because of the wide range of technologies that are considered, it is impossible to present complete details on all of these areas within a reasonable volume. This chapter should therefore be treated as a overview or as a springboard into a more detailed examination of each topic. A list of references is given to facilitate such an effort.

2.1 Aircraft Dynamics

The nonlinear equations of motion for an aircraft are derived using Newton's Second Law of motion. That is, the total sum of all external forces acting on a body must equal the time rate of change of linear momentum and the total sum of all external moments acting on a body must equal the time rate of change of angular momentum. Some critical assumptions necessary to simplify the derivation of the aircraft equations of motion are: the aircraft is a rigid body, the mass of the aircraft remains constant with time, and the earth provides a fixed inertial reference frame [2.1]. Fig. 2.1 shows the body axis forces, moments, angles, velocities, and rotational rates necessary to describe the motion of an aircraft.

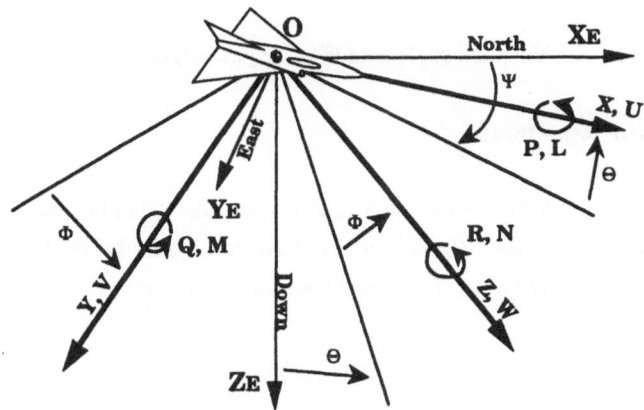

Fig. 2.1 Aircraft Axis System

The nonlinear equations of motion are made up of three translational and three rotational equations. The translational equations of motion are:

$$m[\dot{U} + QW - RV + g \sin \Theta] = X$$
$$m[\dot{V} + RU - PW - g\cos \Theta \sin \Phi] = Y \qquad (2.1)$$
$$m[\dot{W} + PV - QU - g\cos \Theta \cos \Phi] = Z$$

The rotational equations of motion are:

$$\dot{P}I_x - \dot{R}I_{xz} + QR(I_z - I_y) - PQI_{xz} = L$$
$$\dot{Q}I_y + PR(I_x - I_z) - R^2I_{xz} + P^2I_{xz} = M \qquad (2.2)$$
$$\dot{R}I_z - \dot{P}I_{xz} + PQ(I_y - I_x) + QRI_{xz} = N$$

U, V, and W are the translational velocities, P, Q, and R are the rotational rates, m is the aircraft mass, I_x, I_y, I_z, and I_{xz} are the moments of inertia, g is gravity, and X, Y, Z, L, M, and N are the external forces and moments due to the aerodynamics and propulsion. Eqs. (2.1) and (2.2) completely describe the motion of an aircraft. The Euler angles, Θ, Φ, and Ψ, describe the orientation of the aircraft with respect to the Earth.

$$\dot{\Phi} = P + Q \tan \Theta \sin \Phi + R \tan \Theta \cos \Phi$$
$$\dot{\Theta} = Q \cos \Phi - R \sin \Phi \qquad (2.3)$$
$$\dot{\Psi} = \frac{R \cos \Phi}{\cos \Theta} + \frac{Q \sin \Phi}{\cos \Theta}$$

Φ is the roll angle, Θ is the pitch angle, and Ψ is the yaw angle [2.2].

2.1.1 Trimmed Equations

A trimmed condition is a local equilibrium condition at which all of the linear and rotational accelerations in eqs. (2.1) and (2.2) are zero. Let the trimmed flight condition be described as variables with zero subscripts. The general nonlinear trim equations of motion are [2.2]

$$m[Q_0 W_0 - R_0 V_0 + g \sin \Theta_0] = X_0$$
$$m[R_0 U_0 - P_0 W_0 - g\cos \Theta_0 \sin \Phi_0] = Y_0$$
$$m[P_0 V_0 - Q_0 U_0 - g\cos \Theta_0 \cos \Phi_0] = Z_0 \qquad (2.4)$$
$$Q_0 R_0 (I_z - I_y) - P_0 Q_0 I_{xz} = L_0$$
$$P_0 R_0 (I_x - I_z) - R_0^2 I_{xz} + P_0^2 I_{xz} = M_0$$
$$P_0 Q_0 (I_y - I_x) + Q_0 R_0 I_{xz} = N_0$$

While these equations allow for solutions with non-zero rotational rates and side velocity, additional assumptions are usually applied to further simplify the trim solution. The most common of these is straight and level, coordinated flight. In this case eq. (2.4) simplifies to

$$mg \sin \Theta_0 = X_0$$
$$- mg \cos \Theta_0 \sin \Phi_0 = Y_0$$
$$- mg \cos \Theta_0 \cos \Phi_0 = Z_0 \qquad (2.5)$$
$$0 = L_0$$
$$0 = M_0$$
$$0 = N_0$$

That is, the available aerodynamic control surfaces, throttle controls, etc. must be set to cancel the gravitational and aerodynamic forces and moments on the aircraft.

2.1.2 Longitudinal Linear Equations of Motion

The dynamics of a rigid aircraft are described by the six simultaneous nonlinear equations as shown in eqs. (2.1) and (2.2). These equations can be programmed and digitally integrated by a computer to simulate aircraft motion. Because most analysis and design tools require a linear representation of a system, it is useful to make some assumptions to linearize the aircraft equations of motion. The first step is to break the six equations into two sets of simultaneous equations, three longitudinal and three lateral/directional [2.1].

To develop the longitudinal linear equations of motion, it is assumed that the aircraft is in straight and level unaccelerated flight. The only disturbances on the system considered are external forces X and Z, and external moments M. These disturbances on the equations of motion do not create any sideforce, Y, or any rolling moment, L, or yawing moment, N. Roll rate, yaw rate, and side velocity remain undisturbed so three of the equations can be neglected. The remaining equations are simplified because $V = P = R = \Phi = 0$.

$$m[\dot{U} + QW + g \sin \Theta] = X$$
$$m[\dot{W} - QU - g\cos \Theta] = Z \qquad (2.6)$$
$$\dot{Q}I_y = M$$

Fig. 2.2 shows the orientation of the longitudinal variables with respect to the aircraft and its total velocity vector, V_T. The total angle of attack, α_T, is equal to $\sin^{-1}(\frac{W}{V_T})$.

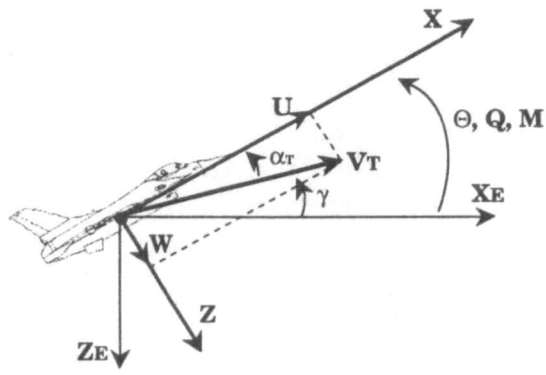

Fig. 2.2 Longitudinal Axis System

By assuming that the aircraft is in equilibrium, the total linear and angular velocities, Euler angles, and the total external forces and moments can be written as the sum of their equilibrium values and perturbational values.

$$
\begin{aligned}
U &= U_0 + u & W &= W_0 + w \\
Q &= Q_0 + q & M &= M_0 + dM \\
X &= X_0 + dX & \Theta &= \Theta_0 + \theta
\end{aligned}
\qquad (2.7)
$$

If only small disturbances about the equilibrium point are considered, the product of the perturbational values can be assumed to be very small and neglected. The angular values between the equilibrium and disturbed conditions are also assumed to be small. By writing the equations in the stabilty axis, W_0 can be set to zero. The longitudinal force and moment equations in (2.6) can now be represented as

$$m[\dot{u} + (g \cos \Theta_0)\theta] = dX$$
$$m[\dot{w} - qU_0 + (g \sin \Theta_0)\theta] = dZ \qquad\qquad (2.8)$$
$$\dot{q}I_y = dM$$

The equations in (2.8) are linear with respect to the perturbational variables. Continued development of the linear longitudinal equations of motion involves expressing expanded representations of the external forces and moments in terms of the changes in them resulting from the perturbations in the linear and angular velocities. In other words the partial derivatives of the forces and moments are taken with respect to the perturbational variables. A detailed development of these equations can be found in [2.1] , [2.2], or [2.3]. The resulting longitudinal linear equations of motion can be expressed in state space form:

$$
\begin{bmatrix} \dot{u} \\ \dot{\alpha} \\ \dot{q} \\ \dot{\theta} \end{bmatrix} =
\begin{bmatrix}
X_u & X_\alpha & 0 & -g \cos \Theta_0 \\
Z_u & Z_\alpha & Z_q & -g/U_0 \sin \Theta_0 \\
M_u & M_\alpha & M_q & 0 \\
0 & 0 & 1 & 0
\end{bmatrix}
\begin{bmatrix} u \\ \alpha \\ q \\ \theta \end{bmatrix} +
\begin{bmatrix}
X_{\delta 1} & X_{\delta 2} & \dots & X_{\delta m} \\
Z_{\delta 1} & Z_{\delta 2} & \dots & Z_{\delta m} \\
M_{\delta 1} & M_{\delta 2} & \dots & M_{\delta m} \\
0 & 0 & \dots & 0
\end{bmatrix}
\begin{bmatrix} \delta_1 \\ \delta_2 \\ \dots \\ \delta_m \end{bmatrix}
$$

$$(2.9)$$

where $\alpha = \alpha_T - \alpha_0$. X_u, Z_α, M_q, etc. are longitudinal stability derivatives and $X_{\delta 1}$, $Z_{\delta 2}$, $M_{\delta m}$, etc. are longitudinal control derivatives. The variables $\delta_1, \delta_2, \dots, \delta_m$ are changes in external control inputs such as elevator, flaps, and thrust vectoring.

Typically, an open loop aircraft with a classical configuration operating in a trimmed condition at a conventional flight condition will exhibit two longitudinal modes of motion: the short period and phugoid. The short period mode is normally fast and oscillatory and takes place at nearly constant speed. It is dominated by angle of attack and pitch rate response. The stability derivative M_α drives the natural frequency of this mode. The phugoid mode is normally slow, oscillatory, and lightly damped and takes place at nearly constant angle of attack [2.3]. For a manual flight control system design problem, the short period is the primary mode of interest because it dominates the aircraft's response to pilot inputs. The phugoid mode is the most important in autopilot designs.

Because the short period mode is virtually decoupled from the speed and pitch angle response of the aircraft, it is possible to reduce the model in eq. (2.9) to a second order short period approximation.

$$\begin{bmatrix} \dot{\alpha} \\ \dot{q} \end{bmatrix} = \begin{bmatrix} Z_\alpha & Z_q \\ M_\alpha & M_q \end{bmatrix} \begin{bmatrix} \alpha \\ q \end{bmatrix} + \begin{bmatrix} Z_{\delta 1} & \cdots & Z_{\delta m} \\ M_{\delta 1} & \cdots & M_{\delta m} \end{bmatrix} \begin{bmatrix} \delta_1 \\ \cdots \\ \delta_m \end{bmatrix} \qquad (2.10)$$

This model provides an accurate measure of the aircraft's transient response to small amplitude inputs over a short time frame (~ 5 seconds). Because the primary focus of this document is manual flight control system design, the phugoid approximation is not described in detail here. The reader is referred to [2.1], [2.2] or [2.3].

2.1.3 Lateral/Directional Linear Equations of Motion

To develop the lateral/directional equations of motion, it is again assumed that the aircraft is in straight and level unaccelerated flight. Only disturbances on the external force Y and external moments L and N are considered [2.1]. The three equations used to derive the decoupled lateral equations of motion are:

$$m[\dot{V} + RU - PW - g\cos\Theta \sin\Phi] = Y$$
$$\dot{P}I_x - \dot{R}I_{xz} + QR(I_z - I_y) - PQI_{xz} = L \qquad (2.11)$$
$$\dot{R}I_z - \dot{P}I_{xz} + PQ(I_y - I_x) + QRI_{xz} = N$$

Fig. 2.3 shows the orientation of the lateral/directional variables with respect to the aircraft and its total velocity vector, V_T. The angle of sideslip, β, is equal to $\sin^{-1}(\frac{V}{V_T})$.

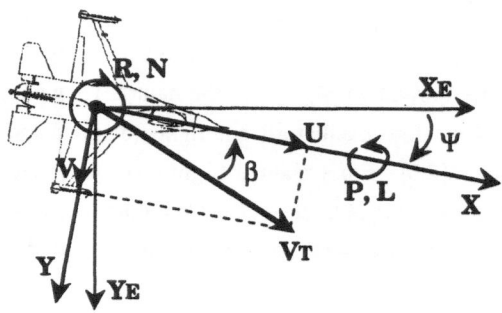

Fig. 2.3 Lateral Axis System

The aircraft is assumed to be in a straight and level equilibrium condition, so the total linear and angular velocities, Euler angles, and the total external forces and moments are represented as the sum of their equilibrium value and perturbational values.

$$
\begin{array}{ll}
P = P_0 + p & R = R_0 + r \\
V = V_0 + v & Y = Y_0 + dY \\
L = L_0 + dL & N = N_0 + dN \\
\Phi = \Phi_0 + \phi & \Psi = \Psi_0 + \psi
\end{array}
\tag{2.12}
$$

Zero total pitch rate, Q, and zero equilibrium side velocity, V_0, can be assumed as a consequence of a decoupled trim condition. The unaccelerated flight condition dictates that equilibrium roll and yaw rates must be zero, $P_0 = R_0 = 0$. Small perturbations about the trim condition are assumed, so the product of these perturbations is neglected [2.1]. In the design of manual flight control systems for fighter aircraft, angles of attack beyond what may be considered small must often be addressed. In order to more clearly express an angle of attack dependency, the translational equation can be rewritten in terms of the stability axis variables sideslip and angle of attack. To maintain decoupling in the equations of motion, angle of attack is assumed constant, $\alpha_T = \alpha_0$. The lateral force and moment equations in (2.12) can now be expressed as

$$
\begin{aligned}
mU_0[\dot{\beta} + r \cos \alpha_0 - p \sin \alpha_0 - g\phi/U_0 \cos \Theta_0] &= dY \\
\dot{p}I_x - \dot{r}I_{xz} &= dL \\
\dot{r}I_z - \dot{p}I_{xz} &= dN
\end{aligned}
\tag{2.13}
$$

The lateral equations in (2.13) are linear with respect to the perturbational variables. They are shown here in a body axis system to show the α_0 dependence in the $\dot{\beta}$ equation. Changes in expanded representations of the external forces and moments resulting from perturbations in the linear and angular lateral velocities must be derived. Detailed derivations are given in [2.1] , [2.2], and [2.3]. The resulting lateral/directional linear equations of motion can be expressed in state space form:

$$
\begin{bmatrix} \dot{\beta} \\ \dot{p} \\ \dot{r} \\ \dot{\phi} \end{bmatrix} = \begin{bmatrix} Y_\beta & \sin \alpha_0 & -\cos \alpha_0 & g/U_0 \cos \Theta_0 \\ L_\beta & L_p & L_r & 0 \\ N_\beta & N_p & N_r & 0 \\ 0 & 1 & 0 & 0 \end{bmatrix} \begin{bmatrix} \beta \\ p \\ r \\ \phi \end{bmatrix} + \begin{bmatrix} Y_{\delta 1} & Y_{\delta 2} & ... & Y_{\delta m} \\ L_{\delta 1} & L_{\delta 2} & ... & L_{\delta m} \\ N_{\delta 1} & N_{\delta 2} & ... & N_{\delta m} \\ 0 & 0 & ... & 0 \end{bmatrix} \begin{bmatrix} \delta_1 \\ \delta_2 \\ ... \\ \delta_m \end{bmatrix}
$$

$$(2.14)$$

Y_β, L_p, N_r, etc. are the lateral stability derivatives and $Y_{\delta 1}$, $L_{\delta 2}$, $M_{\delta m}$, etc. are the lateral control derivatives. Note that Y_p and Y_r are assumed to be negligible.

The lateral/directional equations of motion are usually characterized by three distinct modes of motion: a first order spiral mode, a first order roll mode, and a second order Dutch roll mode. The spiral mode is dominated by bank and heading angle, ϕ and ψ, while sideslip, β, is small. It normally has a long time constant and may be slowly divergent. In manual flight control system designs, the spiral mode is often neglected. Bank and heading angle are trajectory states which should naturally evolve during manuevering flight and therefore should not be regulated by the controller. These states are important in autopilot designs. The roll mode is usually fast and stable. It is dominated by rolling motion and the stability derivative L_p. The Dutch roll mode is dominated by sideslip and yawing motion if the derivative L_β, corresponding to dihedral effect, is small. It is usually oscillatory and lightly damped. The term N_β drives the frequency of this mode, while Y_β and N_r drive the damping [2.3].

2.2   Flying Qualities

Flying qualities include everything that is involved in the safe flight and effective performance of an aircraft, from the pilot's point of view. Military standards for flying qualities provide guidelines for analytical parameters which have been correlated to safety and mission performance. MIL-STD-1797A, Flying Qualities of Piloted Vehicles [2.4], provides guidance for U.S. military, fixed wing aircraft.

In order to interpret requirements, it is necessary to classify an aircraft's mission. Requirements for a light trainer aircraft obviously should not be the same as those for a heavy strategic bomber. An airplane can be put into one of the following categories:

Class I   Small Light Aircraft
Class II   Medium Weight, Low-to-Medium Maneuverability Aircraft

Class III   Large, Heavy, Low-to-Medium Maneuverability Aircraft

Class IV   High-Maneuverability Aircraft

Another important discriminator in flying qualities requirements is flight phase. Categories of flight phases separate flight conditions and tasks into groups which require similar aircraft responses and pilot workloads. These categories are further divided into nonterminal and terminal:

Nonterminal

> **Category A**: Flight phases requiring rapid maneuvering, precision tracking, or precise flight path control.

> **Category B**: Flight phases that are normally accomplished using gradual maneuvers without precision tracking. Accurate flight path control may be required.

Terminal

> **Category C**: Flight phases that are normally accomplished using gradual maneuvers and requiring accurate flight path control.

The qualitative degree of acceptability for handling qualities is given in terms of levels. These levels specify the adequacy of the aircraft response in meeting mission requirements and are based on the Cooper-Harper pilot opinion rating scale [2.4].

> **Level 1 - Satisfactory**: Flying qualities clearly adequate for the mission flight phase. Desired performance achievable with no more than minimal pilot compensation.

> **Level 2 - Acceptable**: Flying qualities adequate to accomplish the mission flight phase with some increase in pilot workload or degradation in mission effectiveness.

> **Level 3 - Controllable**: Flying qualities such that the aircraft can be controlled in the context of the mission flight phase, but with excessive pilot workload or inadequate mission effectiveness.

Most flying quality specifications have been correlated to modal parameters. These modal parameters come from specific low order forms for the transfer functions that describe the aircraft response to pilot inputs. In general, the transfer functions that describe the dynamics of an aircraft include not only the open loop aerodynamics, but also actuator dynamics, controller dynamics, structural modes, sensor dynamics, and so on. These high order transfer functions must be translated into equivalent low order forms so that the modal parameters which define the flying qualities may be identified. These low order forms are called low-order-equivalent-systems (LOES). The LOES is derived by matching the high-order-system (HOS) over a specific frequency range, usually 0.1 rad/s $\leq \omega \leq$ 10 rad/s. The fit is measured by a weighted sum of squares function of the differences in magnitude and phase between the LOES and HOS at n discrete frequencies. This function is given by

$$\frac{20}{n} \sum_{i=1}^{n} [|HOS(j\omega_i)|_{db} - |LOES(j\omega_i)|_{db}]^2 + 0.02[\angle_{deg}HOS(j\omega_i) - \angle_{deg}LOES(j\omega_i)]^2 \quad (2.15)$$

The relative weighting between the gain and phase errors dictates that 1 db of gain mismatch is roughly equal to 7 deg of phase mismatch. Any value of the mismatch function less than 10 is acceptable. Because values greater than 10 may or may not be acceptable, an additional measure of the LOES fit may be used. Fig. 2.4 shows an approximate bound on the maximum unnoticeable dynamics between the LOES and HOS. These bounds represent the level of mismatch that will result in a 1 pilot rating change on the Cooper-Harper scale.   A pilot is most sensitive to dynamics in the region from 1 to 4 rad/s [2.4].

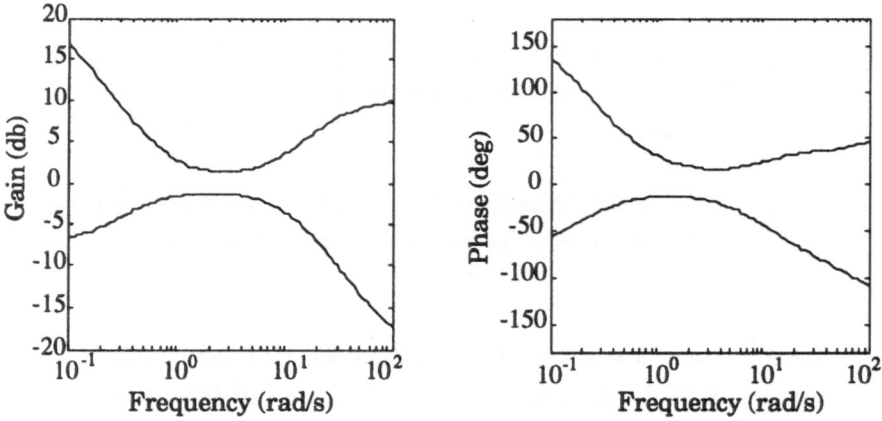

Fig. 2.4 Bounds on Maximum Unnoticeable Added Dynamics

It should be noted that the military standard allows for a good amount of tailoring. Due to the broad possibilities for interpretation, only one possible set of requirements for Class IV aircraft in Category A flight phases is presented here.

2.2.1 Longitudinal Response Requirements

The flying qualities requirements for the pitch axis are determined by the response of the aircraft to a longitudinal input from the pilot, usually a stick force or deflection. The low order modal forms required for the LOES fits are

$$\frac{q(s)}{\delta_p(s)} = \frac{K_\theta s \ (s + 1/T_{\theta_1})(s + 1/T_{\theta_2}) \ e^{-\tau_\theta s}}{(s^2 + 2\zeta_p\omega_p s + \omega_p^2) \ (s^2 + 2\zeta_{sp}\omega_{sp} s + \omega_{sp}^2)} \tag{2.16}$$

and

$$\frac{n_z'(s)}{\delta_p(s)} = \frac{K_n \ (s + 1/T_{n_1}) \ e^{-\tau_n s}}{(s^2 + 2\zeta_p\omega_p s + \omega_p^2) \ (s^2 + 2\zeta_{sp}\omega_{sp} s + \omega_{sp}^2)} \tag{2.17}$$

q is pitch rate, n_z' is the normal acceleration at the instantaneous pitch center of rotation, and δ_p is a pilot pitch stick input.

The primary flying quality measure for the short period mode is the control anticipation parameter (CAP). A value for CAP may be estimated by the expression

$$CAP = \frac{\omega_{sp}^2}{(n/\alpha)} \quad \text{where} \quad (n/\alpha) \approx (V_T/g)(1/T_{\theta_2}) \tag{2.18}$$

V_T is the airspeed along the flight path (ft/s) and g is gravitational acceleration (ft/s^2). The Level 1 and 2 requirements for the estimated CAP parameters are shown in Fig. 2.5.

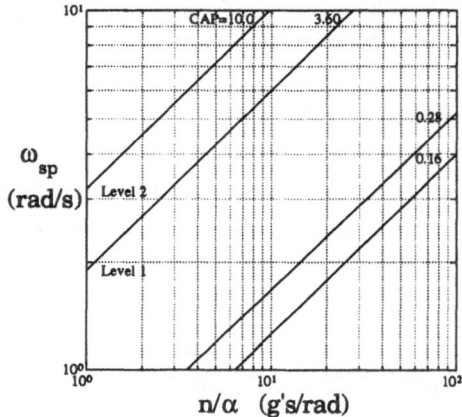

Fig. 2.5 Short Period Frequency Requirements

The requirement specifies that $\omega_{sp} \geq 1.0$ rad/s for Level 1 and $\omega_{sp} \geq 0.6$ rad/s for Level 2. Short period damping, ζ_{sp}, must be between 0.35 and 1.3 for Level 1 and 0.25 and 2.0 for Level 2. Another measure of short period flying qualities is the product of ω_{sp} and T_{θ_2}. Specifications for this product can be used in place of or as a supplement to CAP requirements. Fig. 2.6 shows the values of $\omega_{sp}T_{\theta_2}$ necessary for Level 1 and 2 flying qualities.

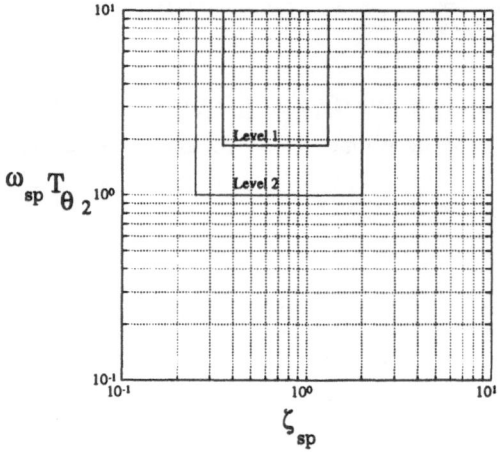

Fig. 2.6 Short Period Damping Requirements

The LOES parameter τ_θ is the equivalent pitch time delay. This value represents lags and time delays in the control system and the phase loss due to the high order dynamics that are not represented in the low order model. Excessive time delay can lead to pilot induced oscillation problems. Table 2.1 shows the requirements for maximum pitch time delay.

Table 2.1 Pitch Time Delay Requirements

Level	Maximum Time Delay
1	0.10 sec
2	0.20 sec
3	0.25 sec

The long term pitch response criteria can be given in terms of the phugoid parameters in the LOES. The stability requirements in Table 2.2 are recommended for any longitudinal mode with a period greater than 15 seconds.

Table 2.2 Phugoid Stability Requirements

Level	Stability Requirement
1	$\zeta_p > 0.04$
2	$\zeta_p > 0.0$
3	$T_d \geq 55$ sec

T_d is the time to double for an unstable oscillation. No aperiodic longitudinal instabilities are acceptable. Often a simple inspection of time response data can be made to check the criteria in Table 2.2. It is then not necessary to find the phugoid parameters through a LOES fit. If this is done, it suffices to use second order LOES to find the short period parameters. The first order numerator and second order denominator terms corresponding to the phugoid mode in eqs. (2.16) and (2.17) can be neglected.

2.2.2 Lateral/Directional Response Requirements

The flying qualities for the lateral/directional axes are determined by the response of the aircraft to a lateral stick input from the pilot. The military standard suggests using roll rate or roll angle as the output for a LOES fit. Because we are concerned with fighter aircraft which may maneuver at elevated angles of attack, an extension to the standard is made to use stability axis roll rate as the output. The low order modal form required for the LOES fit is given by

$$\frac{\dot{\mu}(s)}{\delta_{lat}(s)} = \frac{K_\mu s(s^2 + 2\zeta_\mu \omega_\mu s + \omega_\mu^2) \ e^{-\tau_\mu s}}{(s + 1/T_r)(s + 1/T_s) \ (s^2 + 2\zeta_d \omega_d s + \omega_d^2)} \qquad (2.19)$$

where $\dot{\mu}$ is the stability axis roll rate, $\dot{\mu} = p \cos \alpha + r \sin \alpha$, and δ_{lat} is a lateral stick input. A three step process may be taken in finding a low order fit for eq. (2.19). First, the high order transfer function from lateral stick input to stability axis roll rate is used to find a first order roll mode approximation

$$\frac{\dot{\mu}(s)}{\delta_{lat}(s)} = \frac{K_\mu \ e^{-\tau_\mu s}}{(s + 1/T_r)} \qquad (2.20)$$

Next the high order transfer function from rudder pedal input (δ_{ped}) to sideslip is used to find a second order Dutch roll approximation

$$\frac{\beta(s)}{\delta_{ped}(s)} = \frac{K_\beta(s + 1/T_\beta) \ e^{-\tau_\beta s}}{(s^2 + 2\zeta_d \omega_d s + \omega_d^2)} \qquad (2.21)$$

The roll mode time constant found from the fit of eq. (2.20) and the Dutch roll frequency and damping found from the fit of eq. (2.21) are put into eq. (2.19) and fixed. The fourth order LOES in eq. (2.19) is then fit to the HOS to find the equivalent time delay, numerator poles, and spiral mode time constant.

The general requirement for equivalent roll mode time constant must be less than or equal to 1.0 second for Level 1 and 1.4 seconds for Level 2. These values are conservative for modern fighter aircraft which typically have values of T_r ranging from 1/3 to 1/2 second for conventional flight conditions. At high angles of attack, roll requirements are highly dependent on control power limitations, but typically time constants are slower.

Requirements for Dutch roll frequency and damping are given in Table 2.3. The high Level 1 value for Dutch roll frequency and damping is driven by tracking and pointing requirements in the air-to-air and air-to-ground missions.

Table 2.3 Dutch Roll Frequency and Damping Requirements

Level	Min ζ_d	Min $\zeta_d\omega_d$ (rad/s)	Min ω_d (rad/s)
1	0.4	0.4	1.0
2	0.02	0.05	0.4
3	0	---	0.4

Equivalent roll time delay requirements are identical to the pitch requirements in Table 2.1. The spiral mode typically has a slow first order response. For a manual flight control problem, the spiral mode is generally not a problem, even if it is unstable. The requirements in Table 2.4 define the speed of divergence that is tolerable in an unstable spiral mode.

Table 2.4 Spiral Mode Time to Double Amplitude Requirements

Level	Minimum Time to Double
1	12 sec
2	8 sec
3	4 sec

A number of specific requirements have been described for Class IV aircraft in Category A flight phases. These specifications are derived from the guidelines in MIL-STD-1797A and by no means constitute the complete story on flying qualities. For more detailed information, consult the standard and the references therein [2.4].

2.3   Uncertainty Representation

It is important to remember that any model, no matter how accurate, can only approximate the true behavior of a system. In fact the primary purpose of feedback is to reduce the effect of uncertainty. Robustness is a measure of how tolerant a system is to some level of uncertainty, either structured or unstructured.

Unstructured uncertainty is characterized as additional dynamics that are not represented in the analysis model. Classical single loop gain and phase margins are one way of quantifying robustness to unstructured uncertainty. For multi-loop systems a more general framework is necessary. Unstructured uncertainty may be captured as either an additive or a multiplicative perturbation. Let $G_0(s)$ be the nominal plant and $G(s)$ be the true plant. Δ_a

represents additive uncertainty and Δ_m represents multiplicative uncertainty. Then for an additive uncertainty

$$G(s) = G_0(s) + \Delta_a(s) \ . \tag{2.22}$$

For a multiplicative uncertainty at the plant input

$$G(s) = G_0(s)[I + \Delta_m(s)] \ , \tag{2.23}$$

and for a multiplicative uncertainty at the plant output

$$G(s) = [I + \Delta_m(s)]G_0(s) \ . \tag{2.24}$$

It can be said that an uncertainty is structured if the perturbation corresponds to a parameter or set of parameters whose variation is independent of the rest of the system. If this structure can be captured in an uncertainty model, robustness results will be less conservative. Structured uncertainty can also be captured as either an additive or a multiplicative perturbation. Let l_0 be the nominal parameter value or set of parameter values and l be the true value or set of values. Then using the same nomenclature as above

$$l = l_0 + \Delta_a \tag{2.25}$$

$$l = l_0(1 + \Delta_m) \tag{2.26}$$

The structured singular value (μ) framework described in section 2.4 provides the analysis tools necessary to capture the robustness of a system to any of the above types of uncertainty, structured or unstructured, additive or multiplicative.

2.4   Structured Singular Value Analysis

Structured singular value or μ-analysis is a framework, based on the small gain theorem, in which the robustness of a system can be quantified. Consider the closed loop uncertain system in Fig. 2.7,

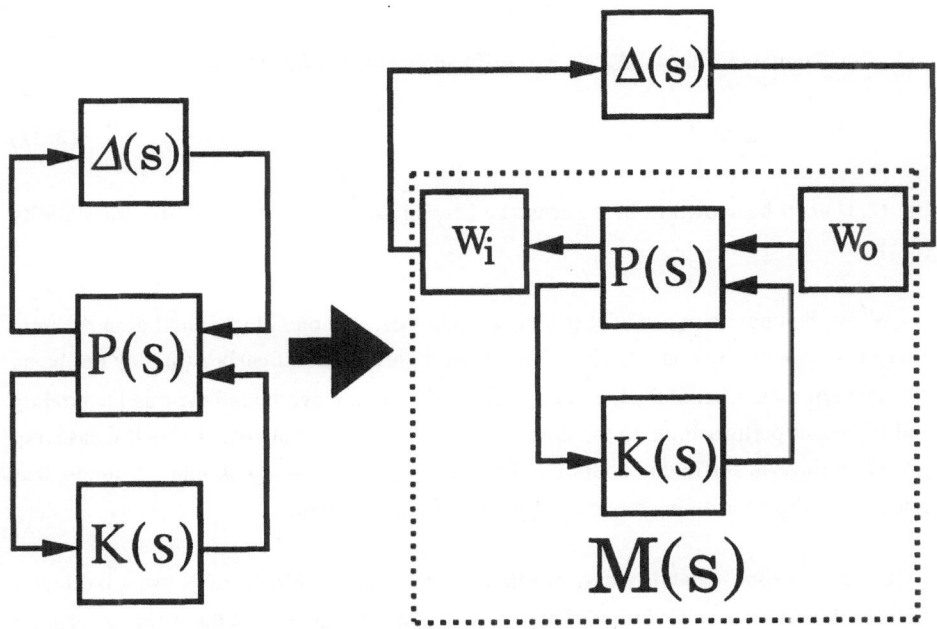

Fig. 2.7 M-Δ Formulation

Any closed loop system with plant, P(s), controller, K(s), and uncertainty, Δ(s), can be rearranged in what is called M-Δ form. An equivalent representation of the uncertainty block is:

$$\Delta(s) = W_o\, \Delta(s)\, W_i, \quad \text{where} \quad \| \Delta(s) \|_\infty \leq 1 . \tag{2.27}$$

The small gain theorem guarantees that if M(s) and Δ(s) are stable, then the uncertain system will remain stable if for all frequencies, $0 \leq \omega \leq \infty$,

$$\bar{\sigma}(\, M(j\omega)\, \Delta(j\omega)\,) \leq 1 , \tag{2.28}$$

which can be equivalently expressed as

$$\| M(s)\ \Delta(s)\ \|_{\infty}\ \leq\ 1. \tag{2.29}$$

An inequality can now be invoked,

$$\| M(s)\ \Delta(s)\ \|_{\infty}\ \leq\ \| M(s)\ \|_{\infty}\ \| \Delta(s)\ \|_{\infty}\ . \tag{2.30}$$

Since it is known that $\| \Delta(s)\ \|_{\infty}\ \leq\ 1$, a sufficient condition for stability is:

$$\| M(s)\ \|_{\infty}\ \leq\ 1. \tag{2.31}$$

Eq. (2.31) can be shown to be a necessary as well as a sufficient condition for stability [2.5].

When the uncertainty model is highly structured, this one block small gain theorem analysis is potentially conservative. The general perturbation matrix $\Delta(s)$ may not be an accurate representation of the true uncertainty, since it may overbound the true uncertainty and represent perturbations to the nominal plant that are unrealistic. In such a case, eq. (2.31) is only a sufficient condition for stability. The analysis can be made less conservative by considering a certain structure for the uncertainty.

Structured robust stability assumes that the perturbation block, $\Delta(s)$, has a particular structure based on specific knowledge of how the uncertainty enters the plant. A general model of the uncertainty is

$$\Delta = \{\ \mathrm{diag}(\delta_1 I_{r_1}, \delta_2 I_{r_2}, ..., \delta_m I_{r_m}, \Delta_1, \Delta_2, ..., \Delta_n)\ |\ \delta_i \in\ C,\ \Delta_j \in\ C^{kj \times kj}\} \tag{2.32}$$

$$B\Delta = \{\ \Delta \in \Delta\ |\ \bar{\sigma}(\Delta) \leq 1\ \} \tag{2.33}$$

where Δ is the set of all possible perturbations. The uncertainty model consists of a diagonal matrix of repeated scalar blocks and full complex uncertainty blocks. A single scalar perturbation is a special case of the repeated block $\delta_m I_{r_m}$, where $r_m = 1$. Unstructured uncertainty is represented as a full complex block.

The structured singular value [2.6] is a measure of robustness to complex perturbations that have a given structure. The structured singular value, μ, of a complex matrix, M, is defined as the inverse of the maximum singular value of the smallest destabilizing perturbation that has the specified structure.

$$\frac{1}{\mu(M)} = \min_{\Delta \subset B\underline{\Delta}} \{ \ \bar{\sigma} \ (\Delta) \mid \det(I - M\Delta) = 0 \ \} \qquad (2.34)$$

If $M(s)$ is a stable closed loop transfer matrix, and $\mu(M(j\omega))$ is evaluated along the imaginary axis, then $\mu(M(j\omega))$ is a function of frequency that gives the size of the smallest allowable Δ which moves a closed loop pole to the imaginary axis. Since the uncertainty representation is scaled to be less than one, if $\mu(M(j\omega))$ is less than one over all frequencies, then the system is stable for all possible uncertainties in the allowed set, $B\underline{\Delta}$.

While in the general case $\mu(M)$ cannot be calculated exactly, its value can be placed between lower and upper bounds. We can define matrices U and D

$$U = \{ \text{diag}(U_1, U_2, ..., U_{m+n}) \mid U_i^* U_i = I \} \qquad (2.35)$$

$$D = \{ \text{diag}(D_1, D_2, ..., D_m, d_1, d_2, ..., d_n) \mid D_i \in C^{r_i \times r_i}, D_i \in D_i^H > 0, d_i \in \Re_+ \} \qquad (2.36)$$

In [2.6], the following properties of the structured singular value are defined:

(a) $\mu(\alpha M) = |\alpha| \mu(M)$

(b) $\mu(I) = 1$

(c) $\mu(AB) \leq \bar{\sigma}(A)\mu(B)$

(d) $\mu(\Delta) = \bar{\sigma}(\Delta)$ for all $\Delta \in \underline{\Delta}$

(e) $U\Delta \in \underline{\Delta}$ and $\Delta U \in \underline{\Delta}$ for all $\Delta \in \underline{\Delta}$ and $U \in U$

(f) $D\Delta D^{-1} = \Delta$ for all $D \in D$ and $\Delta \in \underline{\Delta}$

(g) $\mu(UM) = \mu(MU) = \mu(M)$ for all $U \in U$

(h) $\mu(DMD^{-1}) = \mu(M)$ for all $D \in D$

(i) $\rho(M) \leq \mu(M) \leq \bar{\sigma}(M)$ where $\rho(M)$ denotes the spectral radius of M

By using the properties (g), (h), and (i), the following bounds on $\mu(M)$ are derived.

$$\max_{U \in U} \rho(UM) \leq \mu(M) \leq \inf_{D \in D} \bar{\sigma}(DMD^{-1}) \qquad (2.37)$$

It can be proven that the lower bound is always equal to $\mu(M)$, but the maximization of $\rho(UM)$ is not convex. Local maxima can occur, making a global solution difficult. The minimization of $\bar{\sigma}(DMD^{-1})$ is convex. While this upper bound is an equality only for $n \leq$

3, computational experience has shown that this bound provides a close estimate to the actual value of $\mu(M)$ [2.6,2.7].

In analyzing the robustness of a closed loop system, it is useful to form an analysis model of the form in Fig. 2.8.

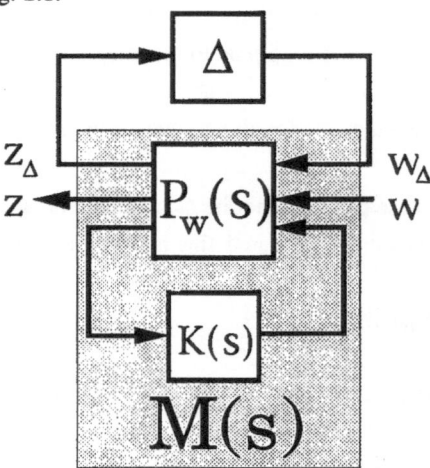

Fig. 2.8 Robust Performance Diagram

P_w is the weighted plant transfer function. $M(s)$ is a set of transfer functions:

$$\begin{bmatrix} z_\Delta \\ z \end{bmatrix} = \begin{bmatrix} M_{11} & M_{12} \\ M_{21} & M_{22} \end{bmatrix} \begin{bmatrix} w_\Delta \\ w \end{bmatrix} \qquad (2.38)$$

It has been shown that robust stability exists if and only if

$$\sup_\omega \mu(M_{11}(j\omega)) \le 1. \qquad (2.39)$$

It is often possible to characterize the performance of a system in terms of a weighted closed loop sensitivity function or model following error. In such a case, a condition for closed loop performance may be written in terms of an infinity norm bound. The transfer function between w and z in Fig. 2.8 represents this weighted sensitivity or model following error. The matrices are scaled such that nominal performance exists if

$$\| M_{22}(s) \|_\infty \le 1 \qquad (2.40)$$

It was shown in [2.7] that by creating a fictitious uncertainty block between w and z, the performance problem can be put into the framework of μ-analysis. This fictitious uncertainty is represented by a full complex block.

It can be said that nominal performance exists if and only if

$$\sup_{\omega} \mu(M_{22}(j\omega)) \leq 1. \tag{2.41}$$

By combining the fictitious performance uncertainty block and the uncertainty blocks for stability robustness analysis, a robust performance analysis model can be formed. If the structured singular value of this combined model is less than unity, then the performance specification of eq. (2.41) will be satisfied in the presence of the bounded uncertainties.

It can be said that robust performance exists if and only if [2.7]

$$\sup_{\omega} \mu(M(j\omega)) \leq 1. \tag{2.42}$$

2.5 Dynamic Inversion

The purpose of dynamic inversion is to develop a feedback control law that linearizes the plant response to commands. In general the nonlinear aircraft dynamics can take the form

$$\dot{x} = f(x,u), \qquad y = Cx \tag{2.43}$$

where x is an n-dimensional state vector, u is a m-dimensional input vector, C is a $p{\times}n$ matrix, and y is a p-dimensional vector of output variables. A transformation is necessary to put the equations in a form from which the inverse dynamics can be constructed. Each controlled output, y_i, is differentiated until an input term from u appears [2.8]. Only m outputs can be controlled independently by the m available inputs, therefore p must equal m. As shown in [2.9], the output equations may now be written in the form,

$$y^{[d]} = \begin{bmatrix} y_1^{[d_1]} \\ y_2^{[d_2]} \\ \vdots \\ y_p^{[d_p]} \end{bmatrix} = h(x) + G(x)u \tag{2.44}$$

where $y_i^{[d_i]}$ represents the d_ith derivative of the output y_i. The inverse dynamics control law can be written as

$$u = G(x)^{-1}(v - h(x)) \tag{2.45}$$

h(x) represents the nonlinear output dynamics and G(x) represents the nonlinear control distribution. The parameter $\vee$ represents the desired linear dynamics of the closed loop system. With the inverse dynamics control law implemented, the closed loop system has the form,

$$y^{[d]} = \vee \qquad (2.46)$$

If the system is observable and $\sum_{i=1}^{p} d_i = n$, then all of the closed loop poles may be placed.

If $\sum_{i=1}^{p} d_i < n$, then closed loop stability cannot be proven. In this case the unobserved dynamics or the *internal dynamics* of dynamic inversion must be checked at local operating points to insure stability [2.8].

2.6 Robust Eigenstructure Assignment

Eigenstructure assignment [2.10] is a technique that uses constant output feedback gains to arbitrarily place the eigenvalues and eigenvectors of the closed loop system. The number of eigenvalues and eigenvectors that can be placed arbitrarily depends upon the number of plant inputs and outputs. Assume a linear time-invariant system in state space form

$$\begin{aligned} \dot{x} &= A x + B u \\ y &= C x \end{aligned} \qquad (2.47)$$

Here, the number of states is n, the number of inputs is m, and the number of outputs is r. The number of eigenvalues that can arbitrarily be placed is equal to the number of outputs, r. The number of elements of the eigenvectors, corresponding to the r eigenvalues that can be placed, is equal to the number of inputs, m.

The eigenstructure assignment problem statement is as follows. Given a set of desired eigenvalues $\lambda_i{}^d$, and eigenvectors $v_i{}^d$, find a real feedback matrix F of dimension m by r such that the closed loop eigenvalues and eigenvectors of $(A + BFC)$ are close to the desired ones. The gain matrix is found by first reordering each of the r desired eigenvectors into m specified and $n-m$ unspecified components using a reordering operator, R_i

$$\{v_i{}^d\} R_i = \begin{bmatrix} l_i{}^d \\ d_i \end{bmatrix} \qquad (2.48)$$

where l_i^d are the m components of the desired eigenvector that are specified, and d_i are the $n-m$ components that are unspecified. Compute the achievable eigenvector for each defined eigenvalue by defining

$$L_i = (\lambda_i^d I - A)^{-1} B \qquad (2.49)$$

and reorder the achievable eigenvector using the corresponding reordering operator

$$\{L_i\}^{Ri} = \begin{bmatrix} \tilde{L}_i \\ D_i \end{bmatrix} \qquad (2.50)$$

where $\tilde{L}_i$ is the specified part of the achievable eigenvector and D_i is the unspecified part. Now, the projection of the desired eigenvector onto the achievable subspace, z_i, is found by minimizing

$$J = \| l_i^d - \tilde{L}_i z_i \|^2 \qquad (2.51)$$

The value of z_i is computed as

$$z_i = (\tilde{L}_i^T \tilde{L}_i)^{-1} \tilde{L}_i^T l_i^d \qquad (2.52)$$

and then the achievable eigenvectors are found as

$$v_i^a = L_i z_i \qquad (2.53)$$

The system is transformed so the B matrix is of the form

$$\begin{aligned} \tilde{B} &= \begin{bmatrix} I \\ 0 \end{bmatrix} = Q^{-1} B \\ \tilde{A} &= Q^{-1} A Q \\ \tilde{C} &= C Q \\ x &= Q \tilde{x} \\ \lambda_i &= \tilde{\lambda}_i \\ v_i^a &= Q \tilde{v}_i^a \end{aligned} \qquad (2.54)$$

where Q is a similarity transformation. The rows of $\tilde{v}_i^a$ and $\tilde{A}$ are partitioned into the first m rows and the last $n-m$ rows

$$\tilde{v}_i{}^a = \begin{bmatrix} \tilde{s}_i \\ \tilde{w}_i \end{bmatrix}, \quad \tilde{A} = \begin{bmatrix} \tilde{A}_1 \\ \tilde{A}_2 \end{bmatrix} \tag{2.55}$$

Define

$$\tilde{S} = [\lambda_1{}^d\tilde{s}_1 \ \lambda_2{}^d\tilde{s}_2 \ .... \ \lambda_r{}^d\tilde{s}_r] \tag{2.56}$$

$$\tilde{V} = [\tilde{v}_1{}^a \ \tilde{v}_2{}^a \ .... \ \tilde{v}_r{}^a] \tag{2.57}$$

Finally, the feedback gain matrix is computed using:

$$F = (\tilde{S} - \tilde{A}_1\tilde{V})(\tilde{C}\tilde{V})^{-1} \tag{2.58}$$

2.7 Full Order H∞ Design

The general H∞ output feedback problem can be characterized simply as finding a controller K, if one exists, that internally stabilizes the closed loop system and satisfies the condition $\|T_{zw}\|_\infty < \gamma$. Fig. 2.9 illustrates the problem.

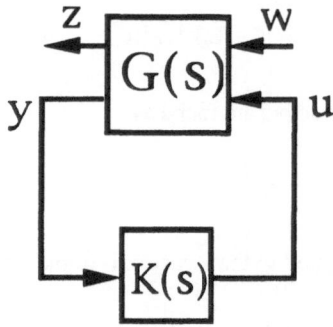

Fig. 2.9 H∞ Problem Formulation

G(s) represents the design model including the plant dynamics and weighting filters.

$$G(s) = \begin{bmatrix} A & B \\ C & D \end{bmatrix} = \begin{bmatrix} A & B_1 & B_2 \\ C_1 & D_{11} & D_{12} \\ C_2 & D_{21} & D_{22} \end{bmatrix} \tag{2.59}$$

The well known state space H_∞ theory [2.11,2.12] gives the equations to produce a stabilizing feedback controller from output vector y to input vector u that minimizes the induced H_∞ norm between w and z. G(s) must satisfy the following conditions:

(1) (A, B_2) is stabilizable and (A, C_2) is detectable

(2) D_{12} is full column rank and D_{21} is full row rank

(3) for all ω, $\begin{bmatrix} A - j\omega I & B_2 \\ C_1 & D_{12} \end{bmatrix}$ has full column rank

(4) for all ω, $\begin{bmatrix} A - j\omega I & B_1 \\ C_2 & D_{21} \end{bmatrix}$ has full row rank

The first condition ensures the existence of a stabilizing controller. Condition (2) can be considered analogous to the nonsingular control penalty and nonsingular sensor noise weight requirements in linear quadratic Gaussian (LQG) control. The column and row rank conditions of (3) and (4) are made to assure that the transfer matrices from w to y and from u to z have no invariant zeros on the imaginary axis. Also assume for convenience that D_{11} and D_{22} are zero, and that the system is scaled so:

$$D_{12}^T [C_1 \ \ D_{12}] = [0 \ \ I]; \qquad \begin{bmatrix} B_1 \\ D_{21} \end{bmatrix} D_{21}^T = \begin{bmatrix} 0 \\ I \end{bmatrix} \qquad (2.60)$$

These last assumptions are made only to simplify the solution, and can be removed. The plant can be manipulated using the approach of [2.13] to transform an arbitrary plant to one that meets these conditions. Define two Hamiltonian matrices and two Riccati equations:

$$H_\infty = \begin{bmatrix} A & \gamma^{-2}B_1B_1^T - B_2B_2^T \\ -C_1^TC_1 & -A^T \end{bmatrix}; \qquad J_\infty = \begin{bmatrix} A^T & \gamma^{-2}C_1^TC_1 - C_2^TC_2 \\ B_1B_1^T & -A \end{bmatrix} \qquad (2.61)$$

$$A^TX_\infty + X_\infty A + X_\infty(\gamma^{-2}B_1B_1^T - B_2B_2^T)X_\infty + C_1^TC_1 = 0 \qquad (2.62)$$

$$AY_\infty + Y_\infty A^T + Y_\infty(\gamma^{-2}C_1^TC_1 - C_2^TC_2)Y_\infty + B_1B_1^T = 0 \qquad (2.63)$$

Under the conditions that the Hamiltonian matrices H_∞ and J_∞ do not have any eigenvalues on the imaginary axis, X_∞ and Y_∞ are positive semidefinite solutions to the two Riccati equations, and the spectral radius of the product of the Riccati solutions, $\rho(X_\infty Y_\infty)$, is less than γ^2, then the controller which satisfies $\|T_{zw}\|_\infty < \gamma$ is given by:

$$\dot{x}_\infty = A_\infty x_\infty \quad - \quad Z_\infty L_\infty y \tag{2.64}$$

$$u = F_\infty x_\infty$$

where:

$$A_\infty = A + \gamma^{-2} B_1 B_1^T X_\infty + B_2 F_\infty + Z_\infty L_\infty C_2$$

$$F_\infty = -B_2^T X_\infty$$

$$L_\infty = -Y_\infty C_2^T \tag{2.65}$$

$$Z_\infty = (I - \gamma^{-2} Y_\infty X_\infty)^{-1}$$

The number of controller states is equal to the order of the design model. While closed loop stability is guaranteed, controller stability is not. Many physically motivated problems do yield a stable compensator.

2.8 Reduced Order Observer Based H_∞ Design

A minimal order H_∞ design algorithm developed in [2.14] considers the linear time-invariant system of Fig. 2.10.

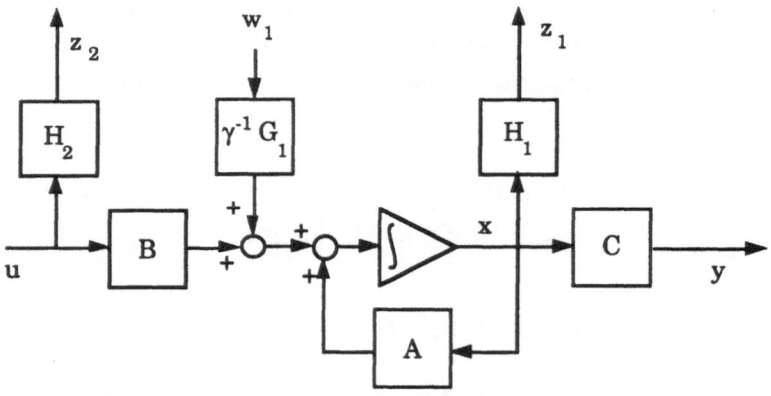

Fig. 2.10 Minimal Order H_∞ Design Model

where (A,B) is stabilizable, (A,C) and (A,H_1) are detectable, and B and H_2 have full column rank.

The minimal-order H_∞ design algorithm generates a controller that stabilizes the closed-loop system and bounds the H_∞-norm of the transfer function from the disturbance (w_1) to the controlled outputs (z_1, z_2) by γ. The controller has the Luenberger observer-based structure shown in Fig. 2.11 and has dimension equal to the number of plant states minus the number of plant measurements.

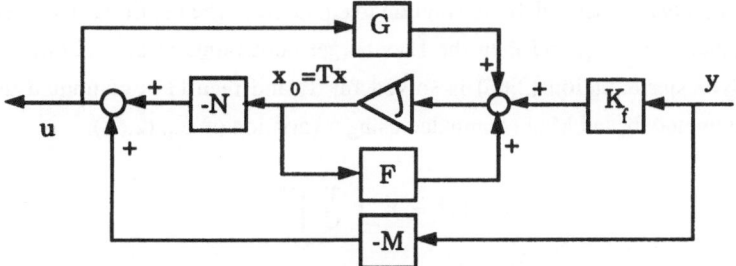

Fig. 2.11 Minimal Order H_∞ Controller Structure

The controller state, x_0, is an estimate of a linear transformation of the plant state, Tx, where T is the transformation matrix. The controller parameters K_f, F, T, M, and N, must satisfy the Luenberger constraints

$$TA - FT = K_f C \tag{2.66}$$

$$G = TB \tag{2.67}$$

$$NT + MC = K_c, \tag{2.68}$$

for any K_c such that $A-BK_c$ is Hurwitz.

According to [2.14], the robust controller parameter K_c is given by

$$K_c = \left[H_2^T H_2\right]^{-1} B^T P_1 \tag{2.69}$$

with $P_1 \geq 0$ satisfying

$$P_1 A + A^T P_1 + H_1^T H_1 + (1+a)\gamma^{-2} P_1 G_1 G_1^T P_1 - P_1 B \left[H_2^T H_2\right]^{-1} B^T P_1 \leq 0 \tag{2.70}$$

subject to the existence of a $P_2 \geq 0$ that satisfies

$$P_2 F + F^T P_2 + N^T H_2^T H_2 N + (1+a^{-1})\gamma^{-2} P_2 T G_1 G_1^T T^T P_2 \leq 0. \tag{2.71}$$

The design parameters a, K_f, and F represent the freedom available to the designer and are chosen such that a is a positive real scalar, F is any stable matrix, and K_f is completely arbitrary. The design tuning parameter a is used to tighten the H_∞ norm bound between the disturbance and the controlled output [2.15]. An initial bound γ is selected that allows the existence of a positive semi-definite solution to eq. (2.70). The controller parameter K_c is now computed using P_1, and then the Luenberger constraints in eqs. (2.66)-(2.68) are solved. Sylvester equation (2.66) is solved for T, and then G is computed using eq. (2.67). Parameters N and M are computed using a variation of eq. (2.68):

$$[N \ M] = K_c \left[\begin{array}{c} T \\ C \end{array} \right]^{-1}. \tag{2.72}$$

Note that the choice of K_f and F provide the design freedom to obtain a T that insures the existence of $\left[\begin{array}{c} T \\ C \end{array} \right]^{-1}$. If there exists a solution to eq. (2.71), a series of γ-reduction iterations is performed on eq. (2.70) to find the smallest bound, γ_{min}, for which positive semi-definite solutions, P_1 and P_2, exist. If no solution exists to eq. (2.71), γ is increased, and the procedure is repeated.

2.9 Structured Singular Value Synthesis

The structured singular value (μ) framework provides a unifying measure which can be used to simultaneously address stability and performance robustness specifications. If μ is less than unity for a properly scaled system, then the specifications are met. It is desirable to be able to address these multiple objectives directly within a design method. μ-synthesis provides for the direct incorporation of robust stability and performance goals into a design by combining H_∞ design with structured singular value analysis [2.16]. The μ-synthesis problem is described by the attempt to find a controller that minimizes an upper bound on the structured singular value,

$$\min_{K} \ \inf_{D \in \mathbf{D}} \ \sup_{\omega} \ \bar{\sigma}(DM(K)D^{-1}). \tag{2.73}$$

M(K) is the weighted closed loop transfer function shown in Fig. 2.8. One approach to this problem is the DK-iteration, it calls for alternately minimizing $\sup \bar{\sigma}(DM(K)D^{-1})$ for either K or D while holding the other constant. First the controller synthesis problem is solved using H_∞ design on the nominal design model, $P_w(s)$. μ-analysis is then performed on the closed loop transfer function M(K), producing values of the D scaling matrices at each frequency. The resulting frequency response data is fit with an invertable, stable, minimum phase transfer function which becomes part of the nominal synthesis structure. With D fixed, the controller synthesis problem is again solved by performing an H_∞ design on the augmented system. The DK-iterations are continued until a satisfactory controller is found or a minimum is reached. Fig. 2.12 shows a flow diagram for the DK-iteration .

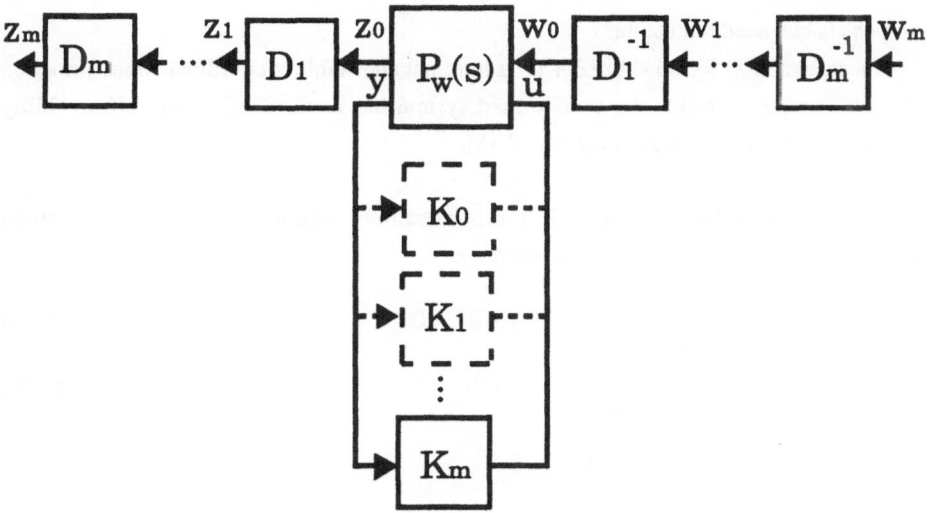

Fig. 2.12 DK-iteration Process

The resulting controller order is the order of the design plant and weighting matrices, as well as the order of the D-scale transfer function fits. With each iteration, the D-scale frequency response data from the previous iteration is combined with the current values, and then the transfer function fit is performed on the combined data. This approach avoids a built-in increase in controller order that would result if at each iteration new D-scale fit transfer functions were augmented into the synthesis model from the previous step. It is important to note that the DK-iteration is not guaranteed to converge to a global minimum, but practical experience has shown that the method works well for a broad class of problems [2.17].

2.10 Balanced Realizations and Truncation

The previous five sections in this chapter have dealt with multivariable compensator design. Unfortunately some of these methods, especially structured singular value synthesis, produce compensators of relatively high order. Limitations in on-board computing power make it desirable to reduce these controllers to an order that can be managed by existing hardware. One method for model or compensator order reduction which is useful for this purpose is balanced truncation.

If a system $G(s) = C(sI - A)^{-1}B + D$ is minimal and stable, a coordinate transformation matrix T exists such that the transformed system has controllability and observability grammians that are equal and diagonal [2.18].

The controllability grammian, P, and observability grammian, Q, are the hermitian solutions to the following Lyapunov equations.

$$AP + PA^T + BB^T = 0 \tag{2.74}$$

$$A^TQ + QA + C^TC = 0 \ . \tag{2.75}$$

The Cholesky factorization of Q is

$$Q = R^TR \ . \tag{2.76}$$

RPR^T is a positive definite matrix, $U\Sigma U^T = RPR^T$ where $U^TU = I$, and Σ is a diagonal matrix of the Hankel singular values in descending order

$$\Sigma = \text{diag}(\sigma_1, \sigma_2, \ldots, \sigma_n), \qquad \sigma_1 \geq \sigma_2 \ldots \geq \sigma_n > 0 \ . \tag{2.77}$$

The transformation which balances the system is

$$T = \Sigma^{-1/2}U^TR \ . \tag{2.78}$$

The controllability and observability grammians of the transformed system are equal and diagonal [2.19].

$$\hat{P} = TPT^T = \Sigma^{-1/2}U^TRPR^TU\Sigma^{-1/2} = \Sigma \tag{2.79}$$

$$\hat{Q} = (T^T)^{-1}QT^{-1} = \Sigma^{-1/2}U^T(R^T)^{-1}(R^TR)R^{-1}U\Sigma^{-1/2} = \Sigma \ . \tag{2.80}$$

By changing the state space coordinates to $z = Tx$, then the transformed system becomes

$$\dot{z} = \hat{A}z + \hat{B}u \qquad y = \hat{C}z + Du ,\tag{2.81}$$

where

$$\hat{A} = TAT^{-1}, \quad \hat{B} = TB, \text{ and } \hat{C} = CT^{-1} .\tag{2.82}$$

Because the Hankel singular values are arranged in descending order, the last elements of the coordinate vector z correspond to the least controllable/observable part of the system. The system in eq. (2.81) can be partitioned into a k dimensional vector z_2 and an n-k dimensional vector z_1.

$$\begin{bmatrix} \dot{z}_1 \\ \dot{z}_2 \end{bmatrix} = \begin{bmatrix} \hat{A}_{11} & \hat{A}_{12} \\ \hat{A}_{21} & \hat{A}_{22} \end{bmatrix} \begin{bmatrix} z_1 \\ z_2 \end{bmatrix} + \begin{bmatrix} \hat{B}_1 \\ \hat{B}_2 \end{bmatrix} u \tag{2.83}$$

$$y = \begin{bmatrix} \hat{C}_1 & \hat{C}_2 \end{bmatrix} \begin{bmatrix} z_1 \\ z_2 \end{bmatrix} + Du \tag{2.84}$$

The controllability and observability grammians of the partitioned system are

$$\hat{P} = \hat{Q} = \begin{bmatrix} \Sigma_{11} & 0 \\ 0 & \Sigma_{22} \end{bmatrix} .\tag{2.85}$$

A reduced order model may now be formed by removing the partition z_2 from the state vector z. This approach is known as balanced truncation. The reduced order system has controllability and observability grammians of Σ_{11}, that is the new system

$$\dot{z}_1 = \hat{A}_{11}z_1 + \hat{B}_1u \qquad y = \hat{C}_1z_1 + Du ,\tag{2.86}$$

satisfies the Lyapunov equations

$$\hat{A}_{11}\Sigma_{11} + \Sigma_{11}\hat{A}_{11}^T + \hat{B}_1\hat{B}_1^T = 0 \tag{2.87}$$

$$\hat{A}_{11}^T\Sigma_{11} + \Sigma_{11}\hat{A}_{11} + \hat{C}_1^T\hat{C}_1 = 0 .\tag{2.88}$$

The reduced order system $G_r = \hat{C}_1(sI - \hat{A}_{11})^{-1}\hat{B}_1 + \hat{D}$ preserves the stability of the original system. The error between the original and reduced order system has an L_∞ error bound given by [2.19]

$$\| G(j\omega) - G_r(j\omega) \|_{L_\infty} \le 2(\sigma_{n-k+1} + \sigma_{n-k+2} + \ldots + \sigma_n) \,. \tag{2.89}$$

The direct feedthrough term of the reduced order system does not affect the grammians, so the $\hat{D}$ term can be adjusted to improve the approximation. To force an exact match at high frequencies, set $G_r(j\omega) = G(j\omega)$ with $\omega = \infty$. Then

$$\hat{D} = D \,. \tag{2.90}$$

To force an exact match in steady state gain, set $G_r(j\omega) = G(j\omega)$ with $\omega = 0$. Then

$$\hat{D} = C(-A)^{-1}B + D \,-\, \hat{C}_1(-\hat{A}_{11})^{-1}\hat{B}_1 \,. \tag{2.91}$$

To approximate a match at a single frequency or set of frequencies, $\omega_1, \omega_2, \ldots \omega_m$, the direct feedthrough term is chosen to solve

$$\min_{\hat{D}} \; \sum_{i=1}^{m} \| G(j\omega_i) - G_r(j\omega_i) \|^2 \,, \tag{2.92}$$

which has the solution

$$\hat{D} = \frac{1}{m} \sum_{i=1}^{m} \mathrm{Re}(C(j\omega_i I - A)^{-1}B + D - \hat{C}_1(j\omega_i I - \hat{A}_{11})^{-1}\hat{B}_1) \,. \tag{2.93}$$

Other variations of balanced trucation include optimal Hankel norm approximations [2.19] and frequency weighted balanced truncation [2.20].

2.11 Conclusions

This chapter provides an overview of the technical details behind the multivariable manual flight control problem. While these tools by themselves do not provide a solution, they represent a pool of knowledge that when combined with engineering experience creates a powerful capability for the design of flight control laws. The next chapter presents a framework for combining these tools to address the issues of gain scheduling, robustness, and performance in a multivariable design.

2.12 References

[2.1] J. H. Blakelock, *Automatic Control of Aircraft and Missiles*, John Wiley & Sons, New York, 1965.

[2.2] D. McRuer, I. Ashkenas, and D. Graham, *Aircraft Dynamics and Automatic Control*, Princeton University Press, Princeton, NJ, 1973.

[2.3] J. Roskam, *Airplane Flight Dynamics and Automatic Flight Controls*, Part 1, Roskam Aviation and Engineering Corporation, Ottawa, KA, 1982.

[2.4] "Military Specification - Flying Qualities of Piloted Vehicles," MIL-STD-1797A, March 1987.

[2.5] J. M. Maciejowski, *Multivariable Feedback Design*, Addison-Wesley, New York, 1989.

[2.6] J. C. Doyle, "Analysis of Feedback Systems with Structured Uncertainties," *IEEE Proceedings*, vol. 129, Part D, No. 6, pp. 242-250, Nov. 1982.

[2.7] J.C. Doyle, J. Wall, and G. Stein, "Performance and Robustness Analysis for Structured Uncertainty," *Proc. 21th IEEE Conf. Decision Contr.*, Dec. 1982.

[2.8] J.E. Slotine and W. Li, *Applied Nonlinear Control*, Prentice Hall, Englewood Cliffs, 1991.

[2.9] S. H. Lane and R.F. Stengel, "Flight Control Design Using Non-linear Inverse Dynamics," *Automatica*, vol. 24, pp. 471-483, 1988.

[2.10] A.N. Andry, E.Y. Shapiro, and J.C. Chung, "Eigenstructure Assignment for Linear Systems," *IEEE Trans. Aerospace and Electronic Systems*, vol. 19, No. 5, pp. 711-728.

[2.11] J. C. Doyle, K. Glover, P. Khargonekar, and B. Francis, "State Space Solutions to Standard H_2 and H_∞ Control Problems," *IEEE Trans. Automat. Contr.*, vol. AC-34, pp. 831-847, Aug. 1989.

[2.12] K. Glover and J. C. Doyle, "State-Space Formulae for all Stabilizing Controllers that Satisfy an H_∞ Norm Bound and Relations to Risk Sensitivity," *Systems and Control Letters*, vol. 11, pp. 167-172, 1988.

[2.13] M.G. Safonov, D.J. Limebeer, and R.Y. Chiang, "Simplifying the H_∞ Theory via Loop Shifting, Matrix Pencil, and Descriptor Concepts," *International Journal of Control*, Vol. 50, No. 6, pp. 2467-2488, 1989.

[2.14] H. H. Yeh, J. L. Rawson, and S. S. Banda, "Robust Control Design with Real-Parameter Uncertainties," *Proc. 1992 American Control Conf.*, Chicago IL, Jun. 1992.

[2.15] J. M. Buffington, H. H. Yeh, and S. S. Banda, "Robust Control Design for an Aircraft Gust Attenuation Problem," *Proc. 31st Conf. on Decision and Control*, Tucson AZ, Dec. 1992.

[2.16] J. C. Doyle, "Structured Uncertainty in Control System Design," *Proc. 24th IEEE Conf. Decision Contr.,* Ft. Lauderdale FL, Dec. 1985.

[2.17]   G. J. Balas, A. K. Packard, J. C. Doyle, K. Glover, and R. S. R. Smith, "Development of Advanced Control Design Software for Researchers and Engineers," *Proc. 1991 American Control Conf.,* Boston MA, June 1991.

[2.18]   B. C. Moore, "Principal Component Analysis in Linear Systems: Controllability, Observability, and Model Reduction," *IEEE Trans. Automat. Contr.,* vol. AC-26, pp. 17-32, Feb. 1981.

[2.19] K. Glover, "All Optimal Hankel-Norm Approximations of Linear Multivariable Systems and Their L_{∞} Error Bounds," *International Journal of Control,* vol. 39, pp. 1115-1193, 1984.

[2.20]   D. Enns, "Model Reduction for Control System Design," Doctoral Dissertation, Stanford University, 1984.

CHAPTER 3

CONTROL DESIGN METHODOLOGY

This section details a design methodology for manual flight control systems. The purpose of this methodology is to derive control laws that achieve desirable robustness and performance properties across a large range of operating conditions. The approach is based on an inner/outer loop control structure. Control redistribution or the allocation of multiple/redundant control effectors is governed by a control selector. The control selector is based on generalized inverses which normalize the control effectiveness with respect to generalized inputs. The gain scheduling problem is addressed by an inner loop compensator which is formulated to equalize the plant dynamics across the flight envelope. This inner loop compensator is a static output feedback or low order controller that is a function of flight parameters such as altitude, Mach number, and angle of attack. The outer loop controller is designed using advanced techniques to achieve performance and robustness goals. This controller should not require gain scheduling because of the equalization performed by the inner loop. These three elements, the control selector, the inner equalization loop, and the outer performance loop, provide a framework for the application of the tools of Chapter 2 to the manual flight control problem.

3.1 Control Selector

The control selector, sometimes referred to as pseudo-controls, has two functions. The first is to normalize control effectiveness by transforming generalized rotational rate commands into actuator position commands. The second is to take advantage of available control redundancy by allowing for control redistribution without changing the linear closed loop performance. The basic idea of the control selector is to redefine the control contribution to the state equation [3.1,3.2],

$$B\delta = B^*\delta^*, \qquad (3.1)$$

where B is the actual control effectiveness matrix, δ is the vector of control deflections, B^* is the generalized control effectiveness matrix, and δ^* is the generalized control vector. The actual control can now be defined in terms of the generalized control,

$$\delta = T\delta^*. \qquad (3.2)$$

The transformation, T, is the control selector. It is defined simply by

$$T = N(BN)^{\#}B^*. \tag{3.3}$$

The operation $(\ )^{\#}$ is a pseudo-inverse and N is a matrix that may be used to combine controls or emphasize/de-emphasize a control channel in the case of redundant effectors. Because the B matrix in eq. (3.3) is a function of flight condition and aircraft state, the control selector is a function of parameters such as Mach number, altitude, and angle of attack.

3.2   Inner Equalization Loop

An inner equalization loop is used to account for the changes in plant dynamics with flight condition. The goal is to make the input/output behavior of the closed loop system uniform for all operating conditions by using a nonlinear static feedback matrix . This inner loop feedback can be derived by scheduling linear compensators or through direct nonlinear methods such as dynamic inversion.

This equalization can be quantified by the concept of relative error. Relative error is defined as

$$\tilde{\Delta}_m := (P - P_0)P_0^{-1} \tag{3.4}$$

where P_0 is the plant used for outer loop design and P is the equalized plant. This idea of relative error gives us a mathematical framework for analyzing the effectiveness of an inner equalization loop. A *Robustness Theorem* by Safonov and Chiang can provide a weak sufficient condition for stability based on relative error [3.3]:

If $\bar{\sigma}(\tilde{\Delta}_m) < 1$ for $\omega < \omega_r$, then the closed loop system will be stable provided that the control bandwidth, ω_b, is less than ω_r.

Thus, if a system with an outer loop controller designed for P_0 is closed loop stable, then closed loop stability should be preserved if P_0 is replaced by P provided that the relative error $\tilde{\Delta}_m$ is sufficiently small.

Explanation: Consider the system with equalized plant, P, outer loop controller, K_{ol}, and relative error, $\tilde{\Delta}_m$. The bandwidth of the control system is defined as the frequency range where the loop transfer function gain is *big* [3.3]. That is:

$$\underline{\sigma}(PK_{ol}) \gg 1 \quad \forall \ \omega < \omega_b \tag{3.5}$$

A sufficient condition for stability is

$$\bar{\sigma}(\tilde{\Delta}_m) \ \bar{\sigma}(PK_{ol}(I + PK_{ol})^{-1}) < 1 \tag{3.6}$$

It follows from eq. (3.5) that for $\omega < \omega_b$,

$$\bar{\sigma}(PK_{ol}(I + PK_{ol})^{-1}) \approx 1 \tag{3.7}$$

So for frequencies where the loop transfer function gain is *big*, a sufficient condition for stability is

$$\bar{\sigma}(\tilde{\Delta}_m) \ < 1 \tag{3.6}$$

This explanation is not a proof in a mathematical sense because of the loose definition of what is considered *big*. Strong claims should not be made about robustness at frequencies where the loop gain is not very large. In addition, the theorem says nothing about what happens at frequencies above the control bandwidth where the relative error may be greater than unity. At these frequencies there is no guarantee that eq. (3.6) will be satisfied. It is also important to remember that eq. (3.6) is only a sufficient condition for stability which can be very conservative. Considering all of these factors, the primary utility of the *Robustness Theorem* for this work is that it provides a framework for quantifying whether or not an inner loop has been successful in equalizing a set of plants. More rigorous post-design analysis is performed to provide well defined bounds for stability and performance robustness.

Fig. 3.1 shows the implementation of the equalization loop using generalized controls.

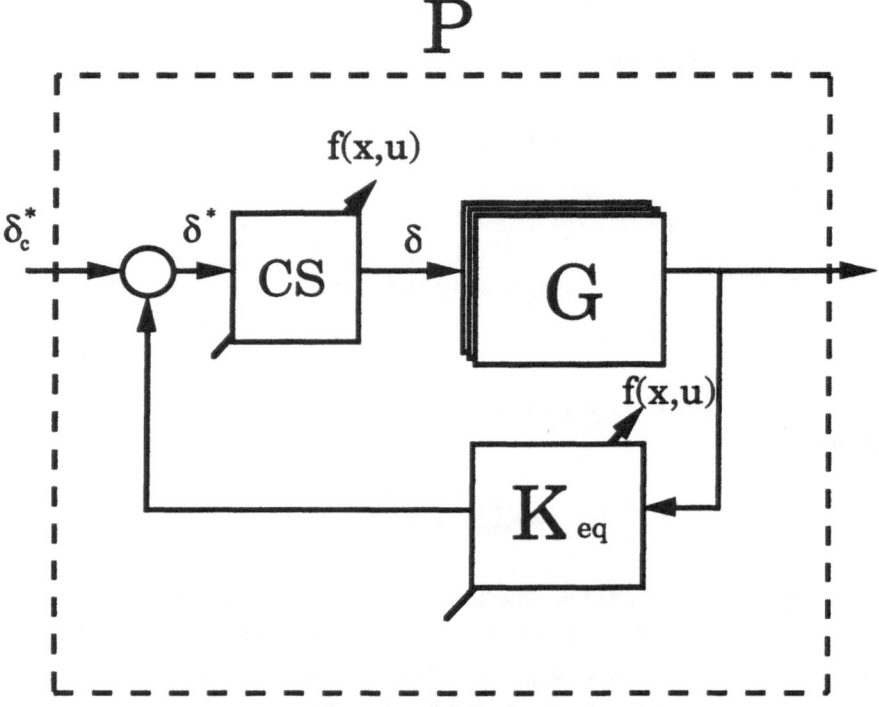

Fig. 3.1. Equalization Loop Implementation

CS is the control selector and K_{eq} is the equalizing controller. With the equalization loop in place, P should approximate a linear system across the flight envelope.

3.3 Outer Robust Performance Loop

The outer robust performance loop uses a fixed output feedback compensator to achieve performance and robustness to parametric uncertainties. The inner equalization loop along with the control selector linearize the aircraft response to generalized commands. By using these generalized commands as control inputs, a single output feedback compensator can satisfy design goals for a broad range of operating conditions. The intractable problem of gain scheduling a high order compensator is eliminated. Any linear design method can now be used to achieve robustness and performance goals. Fig. 3.2 shows the outer loop implementation.

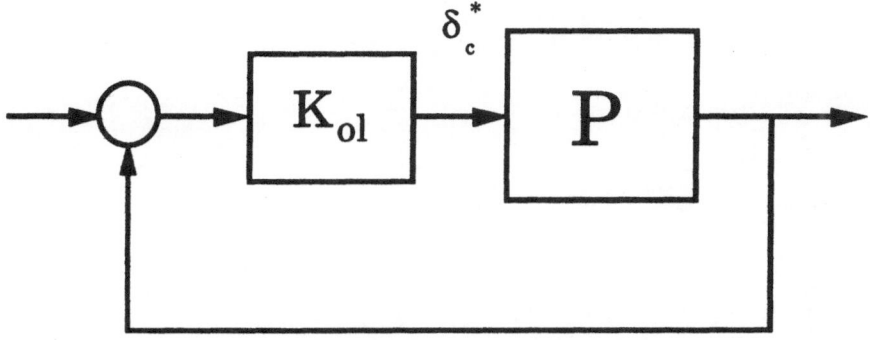

Fig. 3.2 Outer Loop Implementation

3.4 Conclusions

This chapter gives an overview of a methodology for the application of multivariable control theory to multivariable manual flight control laws. The three elements of this approach address specific challenges in the design of flight control laws. The control selector provides for the allocation of multiple and redundant control effectors. The inner equalization loop accounts for the problem of gain scheduling. The outer performance loop provides robust tracking of pilot commands. The next two chapters provide detailed examples of how this methodology can be combined with the tools of Chapter 2 to synthesize designs for realistic fighter aircraft.

3.5 References

[3.1] J. Vincent, "Flight Control Design Considerations for STOVL Powered-Lift Flight," *Proc AIAA/AHS/ASEE Aircraft Design, Systems, and Operations Conf,* Sep. 1990.

[3.2] K. R. Haiges et al., "Robust Control Law Development for Modern Aerospace Vehicles," WL-TR-91-3105, Aug. 1991.

[3.3] M. G. Safonov and R. Y. Chiang, "Model Reduction for Robust Control: A Schur Relative Error Method," *International Journal of Adaptive Control and Signal Processing,* vol. 2, pp. 259-272, 1988.

CHAPTER 4

VISTA F-16 LATERAL/DIRECTIONAL DESIGN

The following sections describe the design and analysis of a full envelope, manual flight control system for the lateral/directional axes of the Variable Stability In-Flight Simulator Test Aircraft (VISTA) F-16 test aircraft. This example shows the utility of the inner/outer loop control structure as well as how various state-of-the-art technologies can be brought together in a comprehensive design. First, the aircraft and specific design requirements are described. Next, the design approach and results are presented, including linear and nonlinear results. Finally, the linear aircraft models and resulting controllers are given in an appendix so that these results may be verified, duplicated, or compared to other approaches.

4.1 Model Description

VISTA is an advanced development program in the Flight Dynamics Directorate of Wright Laboratory. VISTA is a modified F-16 with the capability to simulate advanced aircraft configurations and test advanced flight control concepts. Fig. 4.1 shows a three-view of the F-16 with AIM-9 heat seeking air-to-air missiles on the wingtips. The aircraft has leading and trailing edge flaps, an all moveable horizontal tail, and a single rudder. The pilot controls include a force-feel side stick, rudder pedals, and throttle.

A high fidelity, six degree of freedom, nonlinear simulation model has been developed for the VISTA F-16 vehicle. The VISTA nonlinear model is written as a series of FORTRAN subroutines and is implemented and validated in a generic nonlinear simulation environment. Specific application modules include subroutines describing the equations of motion, actuators, and sensors. The actuation models for all of the control surface deflections include fourth order linear dynamics, hinge moment effects, and position and rate limits. Additional subroutines describe the propulsion system, weight and moments of inertia variations, and the atmosphere. The aerodynamic data exist for a wide range of Mach numbers, altitudes, and angles of attack and sideslip. The high fidelity model allows for the simulation of long duration and large amplitude maneuvers with extreme accuracy. The nonlinear model is used to generate linear models for control law design and to generate nonlinear time histories to evaluate control designs. A diagram of the VISTA F-16 nonlinear model structure is shown in Fig. 4.2, where CMD represents the pilot command.

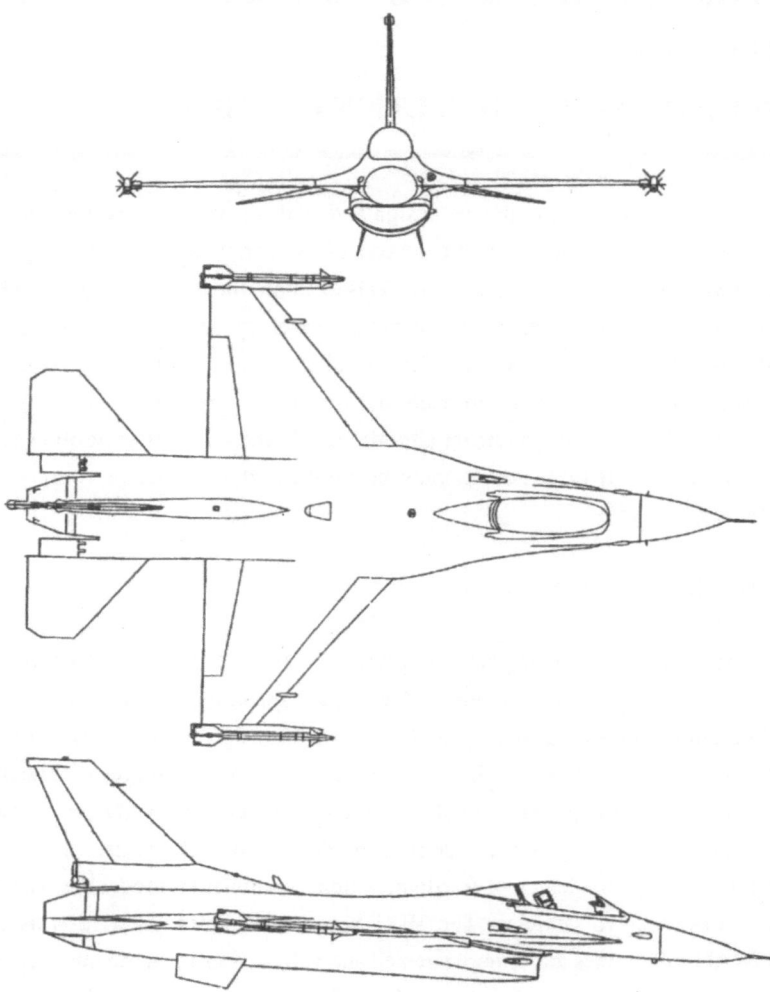

Fig. 4.1 F-16 Three-View

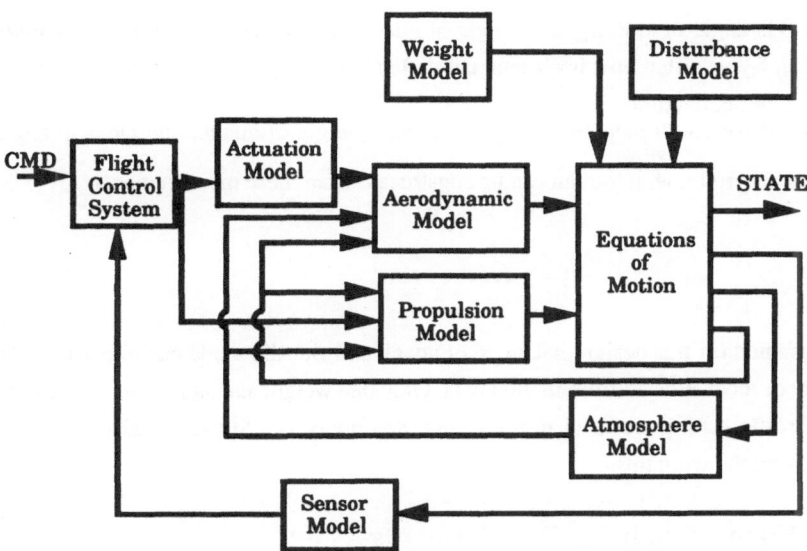

Fig. 4.2 Nonlinear Model Block Diagram

A trim solution can be derived for the aircraft at a specific Mach number and altitude at which all linear and angular accelerations are zero. The solution to eq. (2.4) is found through a numerical routine implemented in the generic simulation environment. Once a trim condition is established for the nonlinear aircraft model, a linear model can be generated to capture the perturbational dynamics around the equilibrium point. Linear models are generated for the open loop aircraft using two-sided perturbations to evaluate the necessary partial derivatives. The linearization procedure yields a tenth order model of open loop aircraft dynamics. The model states are transformed into conventional aircraft states: Euler angles, body axis rotational rates, forward speed, angles of attack and sideslip, and altitude. For a wings level trimmed condition, the linear model consists of decoupled fifth order longitudinal and lateral/directional models. The lateral/directional linear model is reduced to third order by removing the trajectory states: the neutrally stable heading angle and the roll angle. The lateral/directional linear model becomes:

$$\dot{\beta} = Y_\beta\beta + \sin\alpha\, p - \cos\alpha\, r + Y_{\delta DT}\delta_{DT} + Y_{\delta DF}\delta_{DF} + Y_{\delta R}\delta_R$$

$$\dot{p} = L_\beta\beta + L_p p + L_r r + L_{\delta DT}\delta_{DT} + L_{\delta DF}\delta_{DF} + L_{\delta R}\delta_R \tag{4.1}$$

$$\dot{r} = N_\beta\beta + N_p p + N_r r + N_{\delta DT}\delta_{DT} + N_{\delta DF}\delta_{DF} + N_{\delta R}\delta_R \quad ;$$

where β is angle of sideslip, α is angle of attack, p is body axis roll rate, r is body axis yaw rate, δ_{DT} is differential horizontal tail deflection, δ_{DF} is differential flap deflection, and δ_R is rudder deflection.

The measured outputs are roll and yaw rates, angle of attack, and sideslip angle. A stability axis roll rate, $\dot{\mu}$, output can be constructed from these measurements

$$\dot{\mu} = p \cos\alpha + r \sin\alpha . \tag{4.2}$$

It is assumed for this design that angle of attack and sideslip angle can either be measured directly or reconstructed from inertial data. Only one weight and store configuration and the trim throttle setting are used in the design and analysis. Stores consist of two AIM-9 missiles on the wing tips.

4.2   Flying Qualities Requirements

A high order transfer function which includes controller dynamics and actuator dynamics is approximated across a frequency range of interest by a low order fit. The parameters of the low order equivalent system are used to determine whether the aircraft response will be acceptable to a pilot. The following requirements, taken from section 2.2, must be met for Level 1 flying qualities: Dutch roll frequency , $\omega_D \geq 1$ rad/s, Dutch roll damping, $\zeta_D \geq$ 0.4, roll mode time constant, $T_R \leq 1.0$ sec, spiral mode stable or time to double > 12 sec, and equivalent time delay, $\tau_\mu \leq 0.10$ sec. For high angle of attack flight the criteria for Dutch roll frequency is relaxed to $\omega_D \geq 0.7$ rad/s and roll mode time constant is relaxed to $T_R \leq 1.5$ sec.

4.3   Control Selector Design

The generalized inputs for this design are body axis roll acceleration command and body axis yaw acceleration command. The true control inputs are asymmetric horizontal tail, asymmetric flaps, and rudder. The corresponding control effectiveness matrices are:

$$B = \begin{bmatrix} Y_{\delta DT} & Y_{\delta DF} & Y_{\delta R} \\ L_{\delta DT} & L_{\delta DF} & L_{\delta R} \\ N_{\delta DT} & N_{\delta DF} & N_{\delta R} \end{bmatrix} \tag{4.3}$$

$$B^* = \begin{bmatrix} 0 & 0 \\ 1 & 0 \\ 0 & 1 \end{bmatrix} . \tag{4.4}$$

It is important to note that the pseudo-inverse in the control selector, eq. (3.3), does not explicitly account for limitations in control deflection. Undesirable results are therefore possible if the control selector transformations are applied naively. One example is canceling roll inputs from the asymmetric flaps and asymmetric horizontal tail that together achieve a very small increment in yaw. Such cases can cause unreasonable control deflections. If N in eq. (3.3) is chosen to be identity in the transformation equations, this is exactly what happens for the VISTA F-16. This behavior can easily be prevented by combining the asymmetric flaps and asymmetric horizontal tail into a single control effector which we will call aileron (δ_A).

$$\begin{bmatrix} \delta_{DT} \\ \delta_{DF} \\ \delta_R \end{bmatrix} = N \begin{bmatrix} \delta_A \\ \delta_R \end{bmatrix} . \tag{4.5}$$

The matrix N fixes the proportion between asymmetric horizontal tail and asymmetric flap commands. Because the primary purpose of the horizontal tail is pitch control, a ratio of 1/4 is used.

$$N = \begin{bmatrix} 0.25 & 0.0 \\ 1.0 & 0.0 \\ 0.0 & 1.0 \end{bmatrix} , \tag{4.6}$$

This choice is based upon engineering judgement and is non-unique. More elegant solutions that maximize control effectiveness are possible. A few possibilities are described in [4.1].

The control selector equation is

$$\delta = T\delta^* \quad \text{where} \quad \delta = \begin{bmatrix} \delta_{DT} \\ \delta_{DF} \\ \delta_R \end{bmatrix} \quad \text{and} \quad \delta^* = \begin{bmatrix} \dot{p}_c \\ \dot{r}_c \end{bmatrix} , \tag{4.7}$$

and the resulting transformation is

$$T = N(BN)^{\#}B^* . \tag{4.8}$$

The control derivatives in eq. (4.3) are stored in a tabular database as a function of Mach, altitude, and angle of attack. The control selector is found at any flight condition by performing a table lookup of these parameters and using eq. (4.8) to calculate T.

4.4 Inner Loop Design

The goal of the inner equalization loop, as described in section 3.2, is to linearize the input/output behavior of the system across the flight envelope. The dominant nonlinearities which must be handled are the changes in the aerodynamic stability and control derivatives with flight condition. In order to achieve our goal of plant equalization, a controller must be found which itself is a function of flight condition.

4.4.1 Inner Loop Formulation

Dynamic inversion is used to develop the inner loop equalization control law. For inner loop synthesis we include only the nonlinearities associated with changes in the aerodynamic parameters as a function of Mach (M), altitude (h), and angle of attack (α). With the control selector implemented and neglecting actuator dynamics, the aircraft lateral equations of motion are:

$$
\begin{bmatrix} \dot{\beta} \\ \dot{p} \\ \dot{r} \end{bmatrix} = \begin{bmatrix} Y_\beta(M,h,\alpha) & \sin\alpha & -\cos\alpha \\ L_\beta(M,h,\alpha) & L_p(M,h,\alpha) & L_r(M,h,\alpha) \\ N_\beta(M,h,\alpha) & N_p(M,h,\alpha) & N_r(M,h,\alpha) \end{bmatrix} \begin{bmatrix} \beta \\ p \\ r \end{bmatrix} + \begin{bmatrix} 0 & 0 \\ 1 & 0 \\ 0 & 1 \end{bmatrix} \begin{bmatrix} \dot{p}_c \\ \dot{r}_c \end{bmatrix}. \qquad (4.9)
$$

Because the control distribution matrix has a column rank of 2, only two outputs may be chosen for the dynamic inversion formulation. The body axis roll and yaw rate states are chosen as the controlled outputs as they represent the dominant fast dynamics of the open loop system. These variables must only be differentiated once for the control to appear in the output equations.

$$
\begin{bmatrix} \dot{p} \\ \dot{r} \end{bmatrix} = \begin{bmatrix} L_\beta(M,h,\alpha) & L_p(M,h,\alpha) & L_r(M,h,\alpha) \\ N_\beta(M,h,\alpha) & N_p(M,h,\alpha) & N_r(M,h,\alpha) \end{bmatrix} \begin{bmatrix} \beta \\ p \\ r \end{bmatrix} + \begin{bmatrix} 1 & 0 \\ 0 & 1 \end{bmatrix} \begin{bmatrix} \dot{p}_c \\ \dot{r}_c \end{bmatrix}. \qquad (4.10)
$$

The inverse dynamics control law can now be written as

$$\begin{bmatrix} \dot{p}_c \\ \dot{r}_c \end{bmatrix} = \left(V - \begin{bmatrix} L_\beta(M,h,\alpha) & L_p(M,h,\alpha) & L_r(M,h,\alpha) \\ N_\beta(M,h,\alpha) & N_p(M,h,\alpha) & N_r(M,h,\alpha) \end{bmatrix} \begin{bmatrix} \beta \\ p \\ r \end{bmatrix} \right), \qquad (4.11)$$

where V is a matrix that represents the desired linear dynamics. Note that the $\dot{\beta}$ portion of the dynamics in eq. (4.9) is not observed by eq. (4.10). The inner loop equalization will therefore not be perfect and the *internal dynamics* of dynamic inversion will show up as inner loop relative error.

The choice of the desired dynamics, V, is a critical step in the formulation of the inner loop. If unreasonable parameters are chosen, the control system will be sensitive to unmodeled dynamics and have a tendency towards actuator rate and position limit saturation. For the VISTA F-16 design, a linear quadratic regulator design is performed at a flight condition which is considered *central*. *Central* simply means that the chosen flight condition is at some dynamic pressure between the minimum and maximum values for the design envelope. This choice is based upon engineering judgment and is made such that the relative error between the open loop *central* model and all other models is reasonably small. The regulated dynamics at this condition become the desired dynamics for the dynamic inversion calculations:

$$A_{nom} = (A_{central} - B^* K_{LQ}) \qquad (4.12)$$

$$V = \begin{bmatrix} A_{nom21} & A_{nom22} & A_{nom23} \\ A_{nom31} & A_{nom32} & A_{nom33} \end{bmatrix} \begin{bmatrix} \beta \\ p \\ r \end{bmatrix}, \qquad (4.13)$$

The inner equalization loop can be represented as a linear state feedback compensator of the form:

$$\begin{bmatrix} \dot{p}_c \\ \dot{r}_c \end{bmatrix} = K_{eq} \begin{bmatrix} \beta \\ p \\ r \end{bmatrix}, \qquad (4.14)$$

where

$$K_{eq} = \begin{bmatrix} A_{nom21}-L_\beta(M,h,\alpha) & A_{nom22}-L_p(M,h,\alpha) & A_{nom23}-L_r(M,h,\alpha) \\ A_{nom31}-N_\beta(M,h,\alpha) & A_{nom32}-N_p(M,h,\alpha) & A_{nom33}-N_r(M,h,\alpha) \end{bmatrix} . \quad \textbf{(4.15)}$$

The aerodynamic parameters in eq. (4.15) are stored in a tabular database. The inner loop feedback is derived at any flight condition by performing a table lookup of these parameters and calculating the gain matrix, K_{eq}. The equalized system is shown in Fig. 4.3.

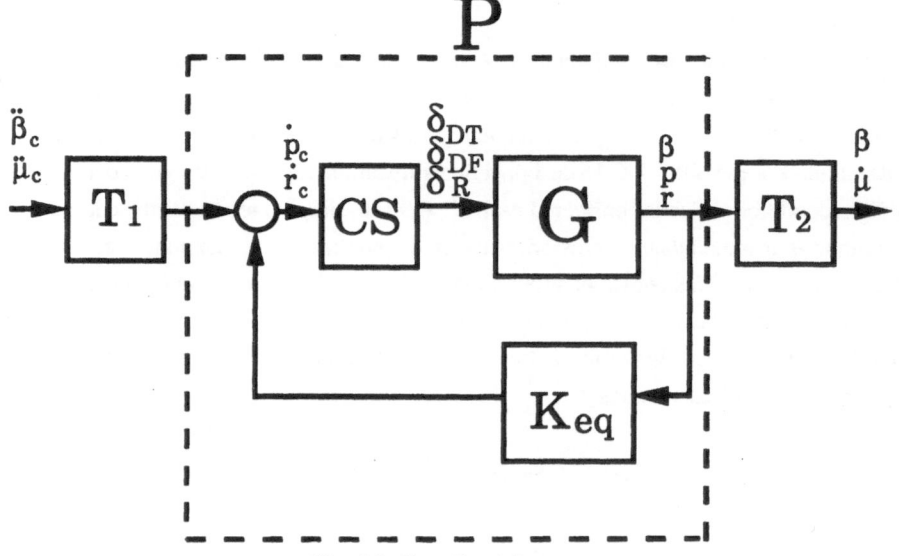

Fig. 4.3 Equalized System

Relative error must be calculated for the system which is to be regulated by the outer loop. The transformation matrices T_1 and T_2 are included so that the outer loop provides regulation in the stability axis rather than in the body axis. The matrices are:

$$T_1 = \begin{bmatrix} \sin\alpha & \cos\alpha \\ -\cos\alpha & \sin\alpha \end{bmatrix} \quad T_2 = \begin{bmatrix} 1 & 0 & 0 \\ 0 & \cos\alpha & \sin\alpha \end{bmatrix} . \quad \textbf{(4.16)}$$

By regulating in the stability axis, the control has a built-in dependence on angle of attack that will alleviate some of the need for scheduling with α. T_1 is a unitary matrix and therefore does not affect singular values.

The two transfer functions of concern for the relative error analysis are $T_2P_0T_1$ and T_2PT_1. P_0 is the system model that is used for the outer loop design. It is the regulated *central* system with third order actuator dynamics included in the inner loop. P is the equalized plant which is formed at each analysis point. The internal dynamics of dynamic inversion and the effects of actuator dynamics are the two primary sources of relative error between these transfer functions.

Two different sets of desired dynamics are used in this application. One is formulated from a *central* model at a low angle of attack (low-α) flight condition and one is formulated from a *central* model at a high angle of attack (high-α) flight condition. These desired dynamics reflect the different robustness and performance requirements at high and low angles of attack.

4.4.2 Low Angle of Attack Inner Loop Results

The *central* point for the low angle of attack design is the trim flight condition at Mach 0.6, altitude 20,000 feet. The dynamic pressure at this point is 245.1 psf. K_{LQ} is the solution to the linear quadratic regulator problem with identity state and control weightings. Details on this design are presented in Appendix 4.

Fig. 4.4 shows the relative error results for four test points. The test points shown in Table 4.1 represent a wide range of operating conditions.

Table 4.1 Test Points for Low-α Linear Analysis

Test Point	Mach	Altitude (ft)	α (deg)	$\bar{q}$ (psf)
Central	0.6	20,000	4.3	245.1
1	0.4	25,000	12.6	87.0
2	0.5	20,000	6.2	170.2
3	0.6	15,000	3.6	301.1
4	0.7	10,000	2.0	499.2

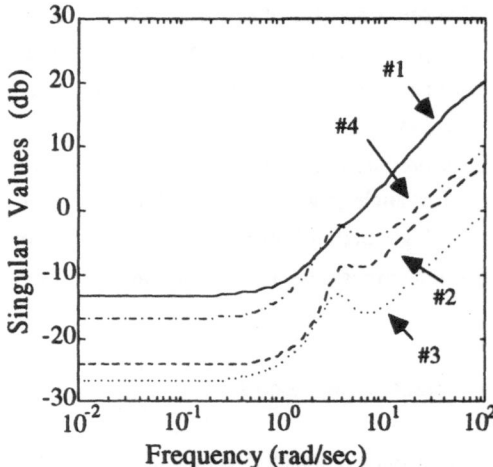

Fig. 4.4 Relative Error Results Low α

While relative error is small (less than 3 db) for the frequency range of interest
(≤ 10 rad/s), it is not less than unity for all conditions tested. This is primarily the result of
interactions between the equalization loop and actuator dynamics. These interactions are
not included in the dynamic inversion calculations. Relative error less than unity is only a
sufficient condition for closed loop stability. Because this is a conservative test, errors that
are small but greater than unity can be tolerated with satisfactory stability and performance
results. The errors shown in Fig. 4.4 are considered small for frequencies of interest.

4.4.3 High Angle of Attack Inner Loop Results

The *central* point for the high angle of attack design is the trim flight condition at Mach
0.3, altitude 25,000 feet. The dynamic pressure at this point is 49.5 psf. K_{LQ} is the
solution to the linear quadratic regulator problem with identity state weighting and a control
weighting of $0.1 \times I_{2 \times 2}$. Details on this design are presented in   Appendix 4.

Fig. 4.5 shows the relative error results for four test points. The test points shown in
Table 4.2 represent a wide range of high-α operating conditions.

Table 4.2 Test Points for High-α Linear Analysis

Test Point	Mach	Altitude (ft)	α (deg)	q̄ (psf)
Central	0.3	25,000	23.3	49.5
1	0.2	10,000	30.0	40.7
2	0.4	20,000	20.0	108.9
3	0.5	15,000	20.0	209.1
4	0.4	25,000	30.0	87.0

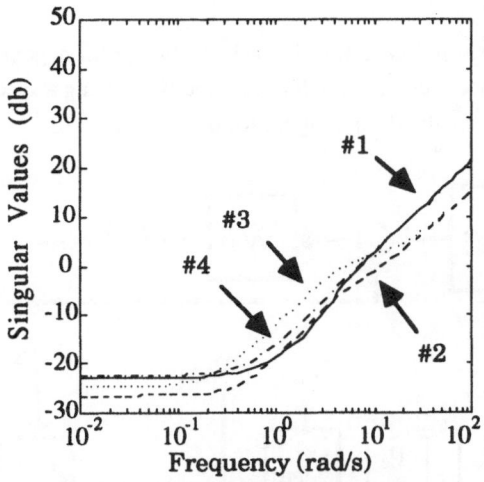

Fig. 4.5 Relative Error Results High-α

The relative error is less than 3 db for frequencies less than 10 rad/s. As in the low-α results, this is primarily the result of interactions between the equalization loop and actuator dynamics. The errors shown in Fig. 4.5 are considered small for frequencies of interest.

4.5   Outer Loop Design

This section describes the formulation and results of the outer loop control design problem. Two separate outer loop designs are performed to reflect the different flying qualities requirements for the low and high angle of attack flight regimes. The formulation of the two designs are the same, only the values of the design parameters differ. Section 4.6 describes how the low-α and high-α designs are blended to create a continuous compensator.

4.5.1 Outer Loop Synthesis

The outer loop compensator is designed to achieve two primary goals, flying qualities and robustness. These objectives can be achieved directly by using a special formulation of μ-synthesis. Fig. 4.6 shows the design model for μ-synthesis.

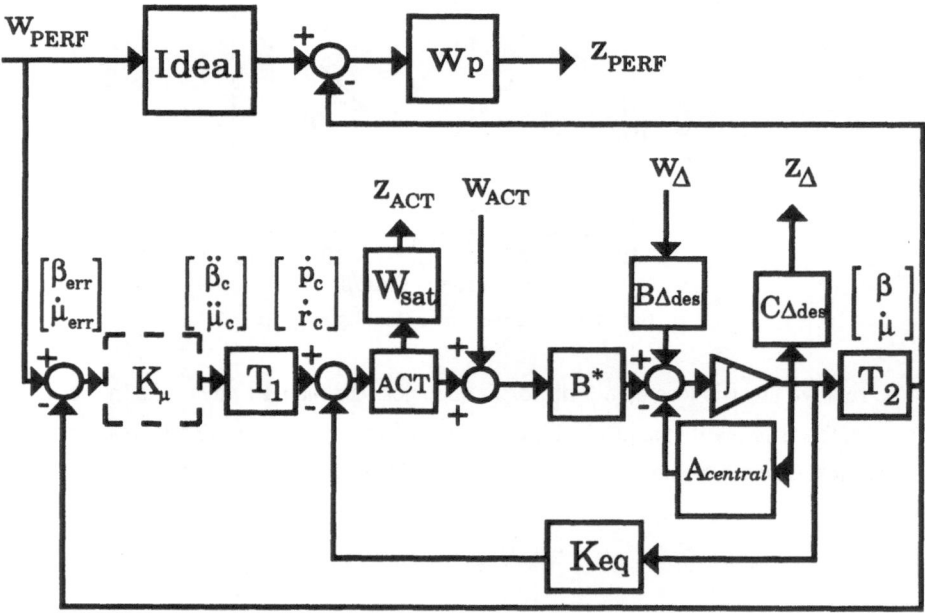

Fig.4.6 Design Model for μ-synthesis

The dynamic compensator that satisfies these goals is found using a DK-iteration based around the regulated *central* plant. The design model can be redrawn to illustrate the μ-

synthesis problem. Fig. 4.7 shows an alternate representation of the design model which parallels the general form shown in Fig. 2.8.

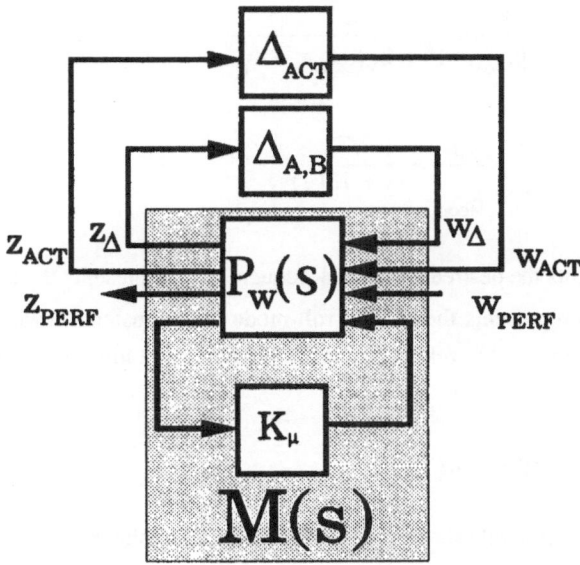

Fig. 4.7 M-Δ Form of Design Model

By using the above form, the VISTA F-16 μ-Synthesis problem can be described as the attempt to solve

$$\min_{K_{\mu\text{-syn}}} \quad \inf_{D \in \mathbf{D}} \quad \sup_{\omega} \; \bar{\sigma}(DM(K_{\mu\text{-syn}})D^{-1}). \qquad (4.17)$$

4.5.1.1 Ideal Model Generation

As described in section 2.2, flying qualities are the primary measures of performance for a manual flight control system. For the VISTA F-16 design, a pilot lateral stick input translates into a stability axis roll rate command, $\dot{\mu}_{com}$, and a pilot pedal input commands a sideslip angle, β_{com}. A low order equivalent system fit of the complementary sensitivity function will therefore drive the flying qualities results for this design. An *ideal model* of the desired aircraft response to pilot inputs can be formulated from the ideal low order equivalent system transfer function parameters. By forcing the complementary sensitivity function to take the frequency response shape of this *ideal model* , flying qualities can be included in the design process [4.2]. The *ideal model* for this design has the form:

$$\frac{\beta}{\beta_{com}} = \frac{\omega_D^2}{s^2 + 2\zeta_D\omega_D s + \omega_D^2} \tag{4.18}$$

$$\frac{\dot{\mu}}{\dot{\mu}_{com}} = \frac{1/T_R}{(s + 1/T_R)} \tag{4.19}$$

where ω_D represents the desired Dutch roll frequency, ζ_D represents the desired Dutch roll damping, and T_R represents the desired roll mode time constant. Two different sets of these parameters are used to reflect the desired dynamics at low and high angle of attack conditions.

4.5.1.2 Performance Weighting

H_∞ optimization in μ-synthesis provides a direct way of minimizing the error between the *ideal model* and the complementary sensitivity function. A weight, W_p, forces this error to be small at frequencies less than 10 rad/s. The performance weight is synthesized such that the error between the ideal model and the complementary sensitivity function is bounded by W_p^{-1} if the design achieves a μ of less than unity. High weightings at low frequencies force the steady state tracking error to be small. The error bound must be tight at frequencies between 1 and 10 rad/s, the region that dominates the transient response of the closed loop system to pilot commands. If the performance error is not reduced adequately in this frequency region, higher order dynamics will show up in the transient response, destroying flying qualities.

4.5.1.3 Actuator Weighting

Actuator dynamics can play a significant role in closed loop performance and robustness. If actuator dynamics are significantly faster than the desired control bandwidth, they may be left out of the compensator synthesis problem. This is often not the case, as it is not with the VISTA F-16. In order to include the dominant actuator effects, a third order approximation of actuator dynamics is included in the outer loop design model. Practical considerations such as rate saturation, unmodeled dynamics, and time delays are included by weighting actuator position, rate, acceleration, and jerk (acceleration rate) through the design weight W_{sat}. Generalized actuator models and static weights are used in the synthesis model to minimize the order of the synthesis model. The generalized actuator model is:

$$\delta^*(s) = \frac{(19.7)(65.0)^2}{(s+19.7)(s^2+2(0.71)(65.0)s+(65.0)^2)} I_{2\times2} \, \delta^*_{com}(s) , \qquad (4.20)$$

where

$$\delta^* = \begin{bmatrix} \dot{p}_c \\ \dot{r}_c \end{bmatrix} . \qquad (4.21)$$

4.5.1.4 Parameter Uncertainty Weighting

Equalization errors and uncertainty in aerodynamic stability derivatives drive the requirements for robustness to parametric uncertainties. Robustness to parameter variations is directly incorporated into the μ-synthesis design model through the weights $B_{\Delta des}$ and $C_{\Delta des}$. As described in section 4.4, the inner loop does not do a perfect job of equalizing the plant. In this design, the internal dynamics of dynamic inversion consist of the A matrix parameters Y_β, $\sin\alpha$, and $\cos\alpha$. By treating these parameters as uncertainties in the μ-synthesis design, the overall control system will be more robust to the relative errors due to equalization. The weighting matrices in the synthesis model are scaled such that the maximum level of uncertainty in the A matrix is represented as:

$$\Delta A = B_{\Delta des} C_{\Delta des}. \qquad (4.22)$$

A trade-off exists between nominal and robust performance in this μ-synthesis design. Flying qualities requirements dictate a tight bound on the allowable error between the ideal model and the complementary sensitivity function. Robust performance and robust stability to parameter uncertainties must be simultaneously achieved in a design that yields μ less than unity. Robust performance, in this case, means that the tight bound on performance error must be met even in the presence of these parameter uncertainties. The tight performance bound was formulated to force nominal performance, but this error bound is too stringent of a requirement in the sense of robust performance. It is impossible to reduce μ to less than unity when a large amount of parameter uncertainty is introduced to the problem. The next logical step seems to be to relax the performance weight. The problem is that this causes the flying qualities specifications to be violated at the nominal condition. No design will be accepted if it cannot meet specs at the nominal condition. The compromise solution is to dramatically reduce the amount of parameter uncertainty introduced in the design model and perform a comprehensive post-design analysis to ensure, that adequate stability and performance robustness has been achieved. Section 4.7 presents the results of this analysis.

4.5.2 Low Angle of Attack Outer Loop Results

The ideal model parameters for the low-α design are $\omega_D = 3.0$ rad/s, $\zeta_D = 0.71$, and $T_R = 0.33$ seconds. The performance weight is:

$$W_p = \frac{(s + 30)}{(s + 0.03)} \, I_{2\times2} \, . \tag{4.23}$$

The generalized actuators are weighted equally. The generalized actuator position weight is 1.5×10^{-3}, the rate weight is 6.0×10^{-4}, the acceleration weight is 6.0×10^{-5}, and the acceleration rate weight is 6.0×10^{-7}. The parameter uncertainty weights used in this design are:

$$B_{\Delta des} = \begin{bmatrix} 0.1 & 0.1 \\ 0 & 0 \\ 0 & 0 \end{bmatrix} \text{ and } C_{\Delta des} = \begin{bmatrix} 0.05 & 0 & 0 \\ 0 & 0.26 & 0 \end{bmatrix}. \tag{4.24}$$

Fig. 4.8 shows the final structured singular value upper and lower bounds that result from the DK-iteration.

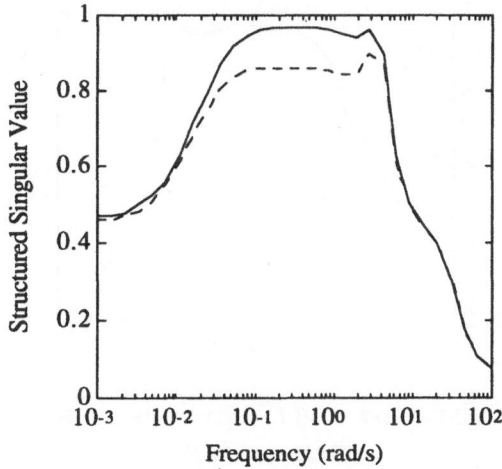

Fig. 4.8 Low-α μ Bounds for Outer Loop Design

The resulting controller is stable and 22nd order. The controller transfer function is shown in pole/zero form in Appendix 4. Using balanced truncation, the outer loop controller is reduced to 15th order. This controller order reduction is an iterative process. First, a balanced realization is found for the full order compensator. The balanced grammians are then examined to determine if there is any natural drop-off in the size of the Hankel singular values. An initial truncation is made and analysis is performed to check if the structured singular values of the closed loop system have been significantly perturbed. A new order is chosen based on this analysis and iterations are continued until the minimal controller order is found that does not deteriorate the μ bounds. It should be noted that a great deal of effort was not put into controller order reduction, and other approaches which could yield a lower order compensator most likely exist. The most promising direction is probably in variations of frequency weighted balanced truncations. Fig. 4.9 shows the μ upper and lower bounds for the system with full and reduced order controllers.

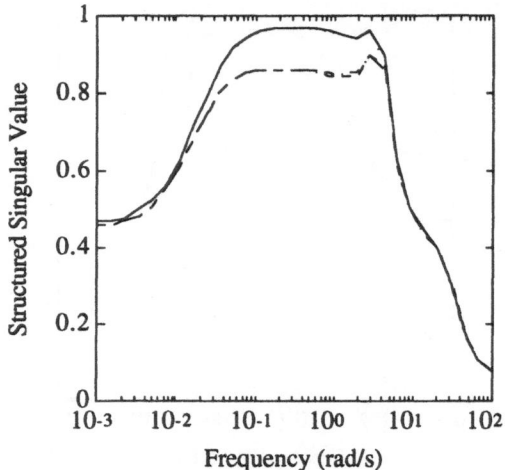

Fig. 4.9 Low-α μ Bounds for Full and Reduced Order Controllers

The reduced order controller transfer function is shown in Appendix 4. The reduced order result matches the full order result very closely, therefore little or no degradation of robustness or performance should be expected.

Fig. 4.10 shows the transfer function between β and β_{com} for the ideal model (dashed line) and for the closed loop system (solid line) at the design condition. The magnitude responses match very closely for frequencies less than 10 rad/s.

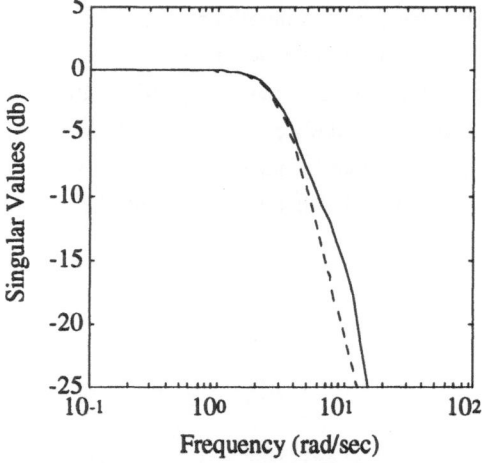

Fig. 4.10 Low-α β/β_{com} for Ideal Model and Outer Loop Design

Fig. 4.11 shows the transfer function between $\dot{\mu}$ and $\dot{\mu}_{com}$ for the ideal model (dashed line) and for the closed loop system (solid line) at the design condition. The compensated system matches the desired first order response very closely for frequencies less than 10 rad/s.

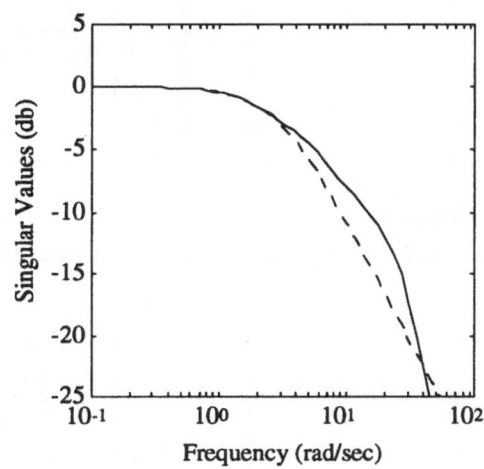

Fig. 4.11 Low-α $\dot{\mu}/\dot{\mu}_{com}$ for Ideal Model and Outer Loop Design

Figures 4.12 and 4.13 show linear responses for unit step commands in sideslip and stability axis roll rate. Solid lines represent the closed loop system response including full order actuator models while the dashed line represent the response of the *ideal model*. Results are shown for the design condition as well as for test conditions 1-4 from Table 4.1.

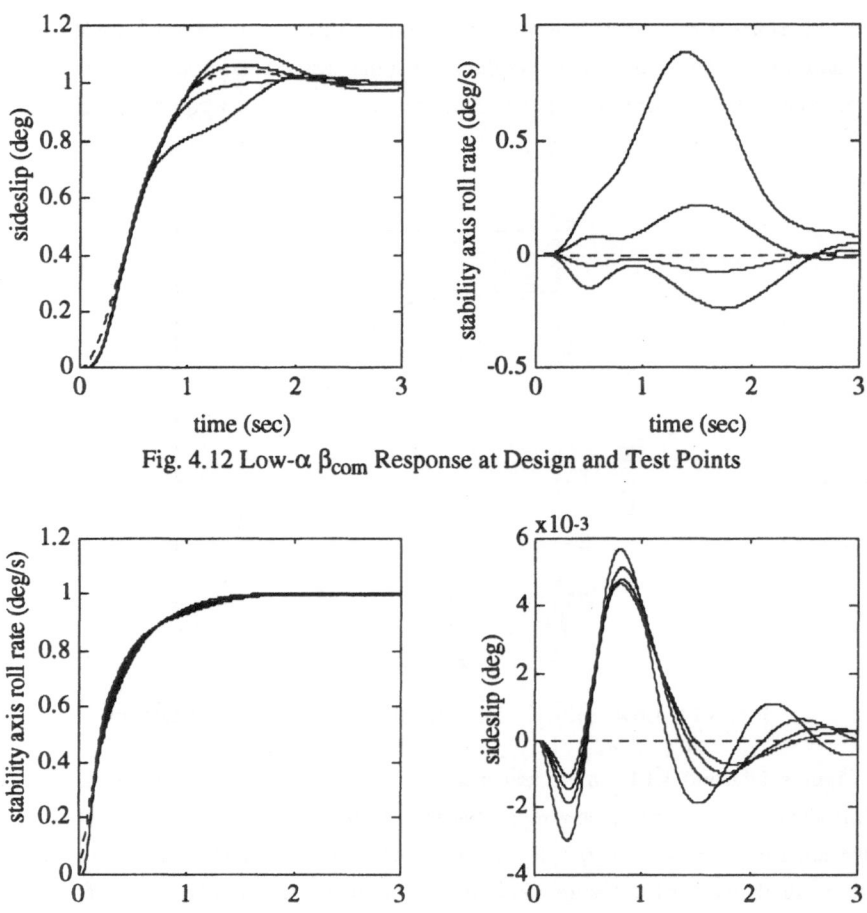

Fig. 4.12 Low-α β_{com} Response at Design and Test Points

Fig. 4.13 Low-α $\dot{\mu}_{com}$ Response at Design and Test Points

 The time response plots show that both tracking and response decoupling are maintained across the low-α design flight envelope, demonstrating the success of the equalization in preserving nominal performance.

4.5.3 High Angle of Attack Outer Loop Results

The ideal model parameters for the high-α design are $\omega_D = 1.0$ rad/s, $\zeta_D = 1.0$, and $T_R = 1.2$ seconds. The performance weight is:

$$W_p = \frac{0.75(s + 30)}{(s + 0.03)} \, I_{2 \times 2} \, . \tag{4.25}$$

The generalized actuators are weighted equally. The generalized actuator position weight is 1.5×10^{-3}, the rate weight is 6.0×10^{-4}, the acceleration weight is 6.0×10^{-5}, and the acceleration rate weight is 6.0×10^{-7}. The parameter uncertainty weights used in this design are:

$$B_{\Delta des} = \begin{bmatrix} 0.1 & 0.1 \\ 0 & 0 \\ 0 & 0 \end{bmatrix} \text{ and } C_{\Delta des} = \begin{bmatrix} 0.05 & 0 & 0 \\ 0 & 0.26 & 0 \end{bmatrix}. \tag{4.26}$$

Fig. 4.14 shows the final structured singular value upper and lower bounds that result from the DK-iteration.

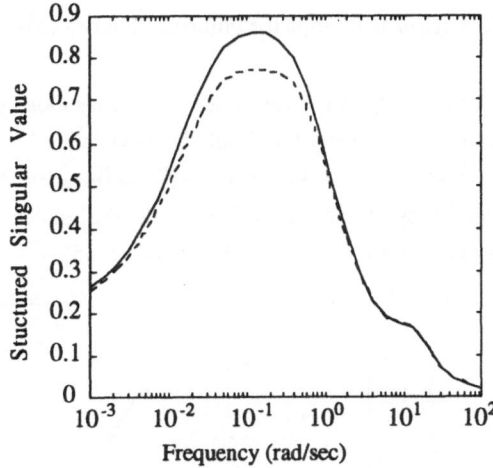

Fig. 4.14 High-α μ Bounds for Outer Loop Design

The resulting controller is stable and 22nd order. Using balanced truncation, the outer loop controller is reduced to 13th order. The full and reduced order controller transfer functions are shown in Appendix 4. Fig. 4.15 shows the μ upper and lower bounds for the system with full and reduced order controllers.

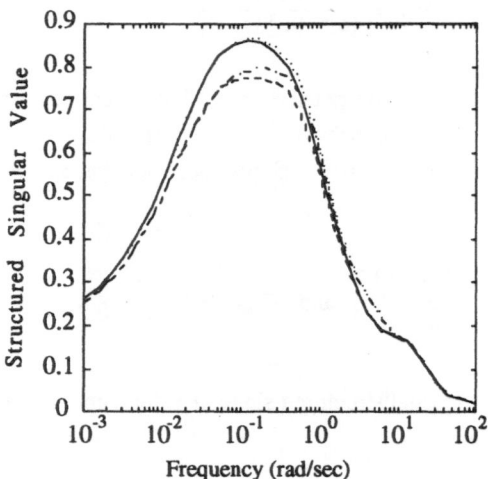

Fig. 4.15   High-α μ Bounds for Full and Reduced Order Controllers

The reduced order result matches the full order result very closely, therefore little or no degradation of robustness or performance should be expected.

Fig. 4.16 shows the transfer function between β and β_{com} for the ideal model (dashed line) and for the closed loop system (solid line) at the design condition. The magnitude responses match very closely for frequencies less than 10 rad/s.

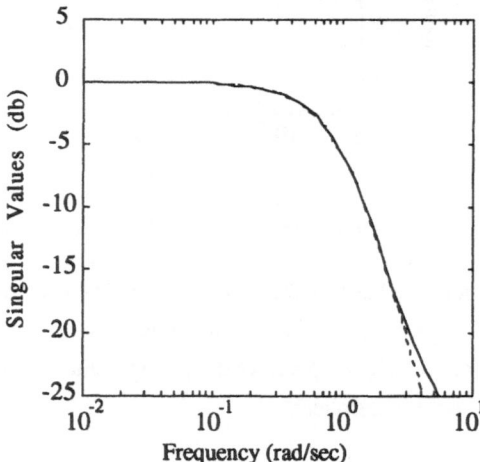

Fig. 4.16 High-α β/β_{com} for Ideal Model and Outer Loop Design

Fig. 4.17 shows the transfer function between $\dot{\mu}$ and $\dot{\mu}_{com}$ for the ideal model (dashed line) and for the closed loop system (solid line) at the design condition. The compensated system matches the desired first order response very closely for frequencies less than 10 rad/s.

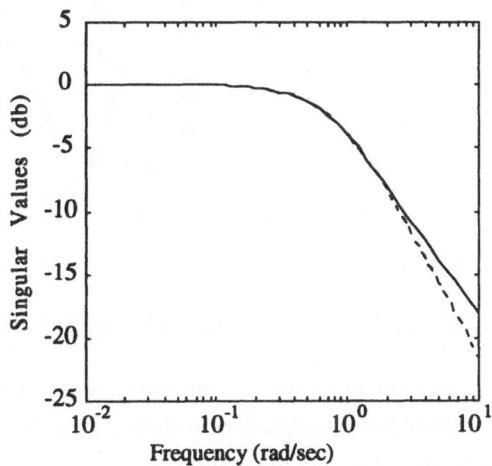

Fig. 4.17 High-α $\dot{\mu}/\dot{\mu}_{com}$ for Ideal Model and Outer Loop Design

Fig. 4.18 and 4.19 show linear responses for unit step commands in sideslip and stability axis roll rate. Solid lines represent the closed loop system response including full order actuator models while the dashed line represent the response of the *ideal model*. Results are shown for the design condition as well as for test conditions 1-4 from Table 4.2.

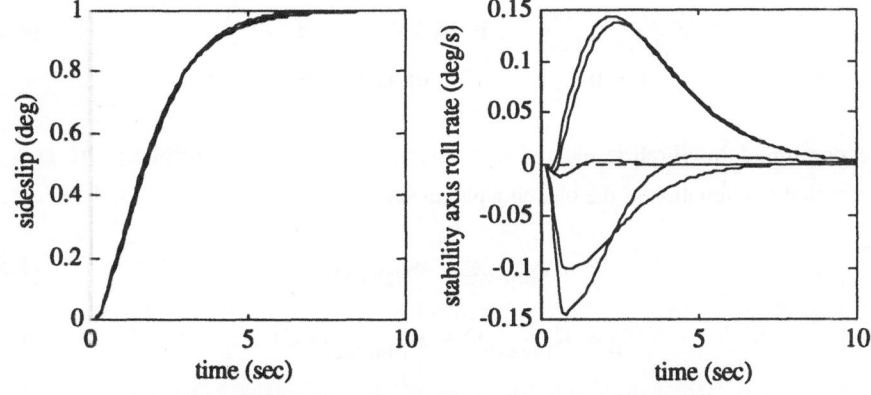

Fig. 4.18 High-α β_{com} Response at Design and Test Points

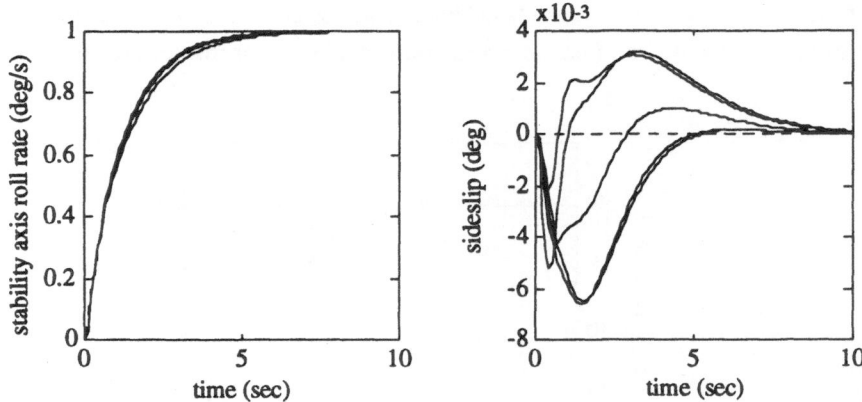

Fig. 4.19 High-α $\dot{\mu}_{com}$ Response at Design and Test Points

The time response plots show that both tracking and response decoupling are maintained across the high-α design flight envelope, demonstrating the success of the equalization in preserving the nominal performance of the *central* design point.

4.6   Controller Implementation

The low-α and high-α controllers must be combined to form a complete full-envelope control system. A blending parameter, C, can be introduced to create a transition region in angle of attack,

$$
\begin{aligned}
C &= 1 && \text{if } \alpha < 12.5 \\
C &= \frac{17.5 - \alpha}{17.5 - 12.5} && \text{if } 12.5 \le \alpha \le 17.5 \\
C &= 0 && \text{if } \alpha > 17.5 \; .
\end{aligned}
\tag{4.27}
$$

The combined low/high-α desired dynamics and outer loop controllers can now be represented as a function of the blending parameter,

$$
\nu = \nu_{low-\alpha}C + \nu_{high-\alpha}(1-C)
\tag{4.28}
$$

$$
K_\mu = K_{\mu_{low-\alpha}}C + K_{\mu_{high-\alpha}}(1-C) \; .
\tag{4.29}
$$

The implementation of the outer loop controller is shown in Fig. 4.20. P represents the equalized system including the control selector and equalization feedbacks.

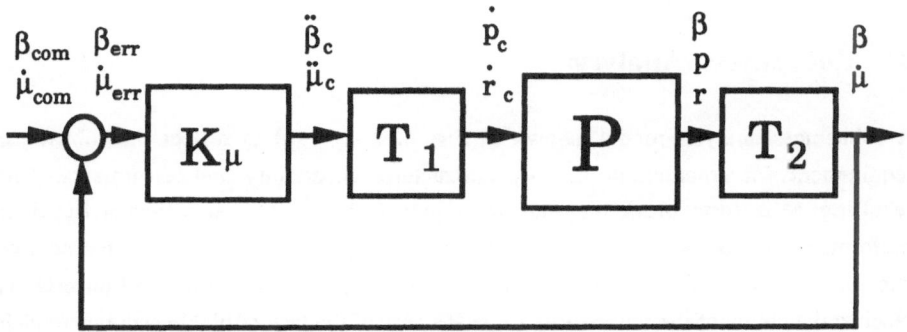

Fig. 4.20 Outer Loop Implementation

The outer loop compensator accepts inputs of sideslip and stability axis roll rate error and generates a sideslip acceleration and stability axis roll acceleration command signal.

Fig. 4.21 shows the overall structure of the VISTA F-16 inner/outer loop controller.

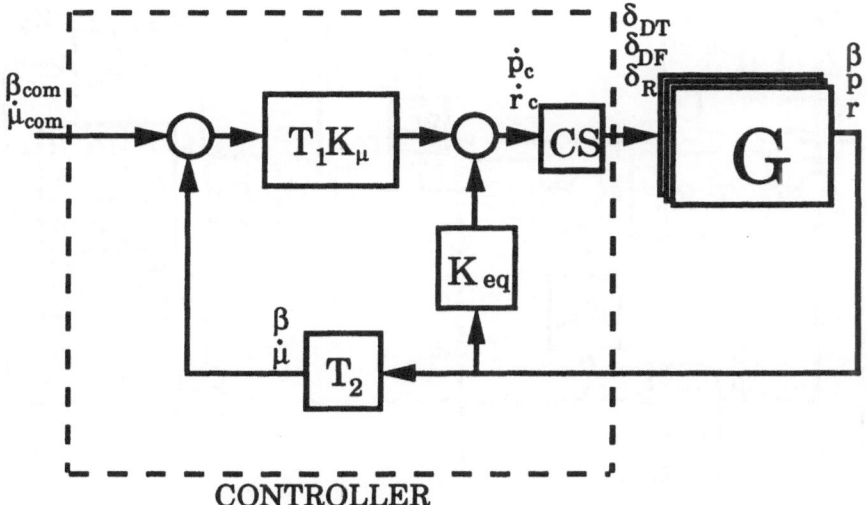

Fig. 4.21 Inner/Outer Loop Controller

The combined controller receives inputs of sideslip, body axis roll rate, body axis yaw rate, sideslip command, and stability axis roll rate command. The controller generates actuator commands for asymmetric flap, asymmetric horizontal tail, and rudder deflections. As described earlier, the equalization gains, $\mathbf{K_{eq}}$, and the control selector, $\mathbf{CS}$, are functions of Mach, altitude, and angle of attack. The outer loop controller, $\mathbf{K_\mu}$, is a dynamic compensator that is a linear function of angle of attack.

4.7 Robustness Analysis

A robustness analysis model, shown in Fig. 4.22, is built to reflect the robustness requirements for structured uncertainty, unstructured uncertainty, and performance. The resulting M-Δ form block diagram has inputs and outputs that reflect a fictitious performance block (w_1 - z_1), a structured uncertainty block (w_2 - z_2), an unstructured uncertainty block at the input to the actuators (w_3 - z_3), and an unstructured uncertainty block at the output of the sensors (w_4 - z_4). Because of the lack of highly accurate models for the different uncertainties, analysis is performed considering each type separately. This approach allows problem areas and trends in robustness to be isolated and identified.

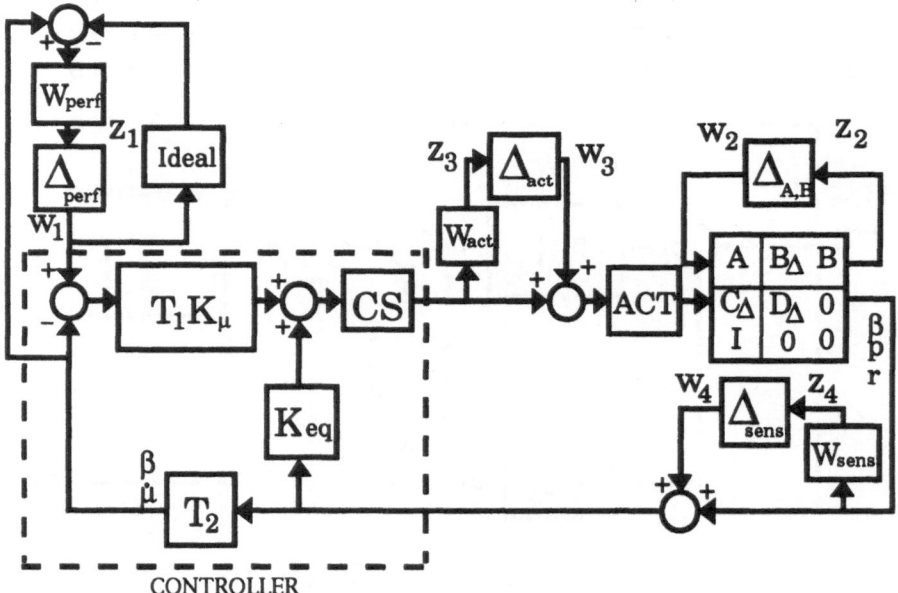

Fig. 4.22 Robustness Analysis Model

4.7.1 Low Angle of Attack Robustness Results

4.7.1.1 Low Angle of Attack Structured Uncertainty

Structured uncertainty requirements for the VISTA F-16 lateral/directional design are driven by perturbations in aerodynamic parameters. Seven stability derivatives and seven control derivatives are identified for robustness analysis. We can write the perturbed state equations as:

$$\dot{x} = (A + \Delta A)x + (B + \Delta B)u \quad , \tag{4.30}$$

where

$$\Delta A = \begin{bmatrix} \Delta Y_\beta & 0 & 0 \\ \Delta L_\beta & \Delta L_p & \Delta L_r \\ \Delta N_\beta & \Delta N_p & \Delta N_r \end{bmatrix} \quad \Delta B = \begin{bmatrix} 0 & 0 & \Delta Y_{\delta R} \\ \Delta L_{\delta DT} & \Delta L_{\delta DF} & \Delta L_{\delta R} \\ \Delta N_{\delta DT} & \Delta N_{\delta DF} & \Delta N_{\delta R} \end{bmatrix} . \tag{4.31}$$

The structure of the uncertainty can be captured by rewriting the perturbed state equations as:

$$\dot{x} = Ax + Bu + B_\Delta \Delta_{A,B} z_2 \tag{4.32}$$

$$z_2 = C_\Delta x + D_\Delta u \quad , \tag{4.33}$$

so when $\bar{\sigma}(\Delta_{A,B}) \leq 1$, the maximum parameter uncertainty in the system matrices is represented as:

$$\Delta A = B_\Delta C_\Delta \quad \Delta B = B_\Delta D_\Delta \quad . \tag{4.34}$$

The level of parameter uncertainty is captured by the matrices:

$$B_\Delta = \begin{bmatrix} 1 & 0 & 0 & 0 & 0 & 0 & 0 & 0 & 0 & 0 & 0 & 1 & 0 & 0 \\ 0 & 1 & 0 & 1 & 0 & 1 & 0 & 1 & 0 & 1 & 0 & 0 & 1 & 0 \\ 0 & 0 & 1 & 0 & 1 & 0 & 1 & 0 & 1 & 0 & 1 & 0 & 0 & 1 \end{bmatrix} \tag{4.35}$$

$$
C_\Delta =
\begin{bmatrix}
\Delta Y_\beta & 0 & 0 \\
\Delta L_\beta & 0 & 0 \\
\Delta N_\beta & 0 & 0 \\
0 & \Delta L_p & 0 \\
0 & \Delta N_p & 0 \\
0 & 0 & \Delta L_r \\
0 & 0 & \Delta N_r \\
& 0_{7\times3} &
\end{bmatrix}
\qquad (4.36)
$$

$$
D_\Delta =
\begin{bmatrix}
& 0_{7\times3} & \\
\Delta L_{\delta DT} & 0 & 0 \\
\Delta N_{\delta DT} & 0 & 0 \\
0 & \Delta L_{\delta DF} & 0 \\
0 & \Delta N_{\delta DF} & 0 \\
0 & 0 & \Delta Y_{\delta R} \\
0 & 0 & \Delta L_{\delta R} \\
0 & 0 & \Delta N_{\delta R}
\end{bmatrix}
. \qquad (4.37)
$$

The level of uncertainty for a parameter at each operating condition is captured as a percentage of its nominal value. The uncertainty levels used for stability robustness analysis are shown in Table 4.3. The values are based upon the accuracy of the VISTA F-16 aerodynamic database and the uncertainty database described in [4.3].

Table 4.3 Structured Uncertainty Levels

stability derivatives	control derivatives
$\Delta Y_\beta = 0.15\, Y_\beta$	$\Delta Y_{\delta R} = 0.15\, Y_{\delta R}$
$\Delta L_\beta = 0.10\, L_\beta$	$\Delta L_{\delta DT} = 0.15\, L_{\delta DT}$
$\Delta L_p = 0.30\, L_p$	$\Delta L_{\delta DF} = 0.10\, L_{\delta DF}$
$\Delta L_r = 0.20\, L_r$	$\Delta L_{\delta R} = 0.40\, L_{\delta R}$
$\Delta N_\beta = 0.30\, N_\beta$	$\Delta N_{\delta DT} = 0.15\, N_{\delta DT}$
$\Delta N_p = 0.50\, N_p$	$\Delta N_{\delta DF} = 0.20\, N_{\delta DF}$
$\Delta N_r = 0.15\, N_r$	$\Delta N_{\delta R} = 0.15\, N_{\delta R}$

Stability robustness to structured uncertainties is tested for the *central* plant and the four test points described in Table 4.1. Fig. 4.23 shows the resulting upper and lower bounds on the structured singular value.

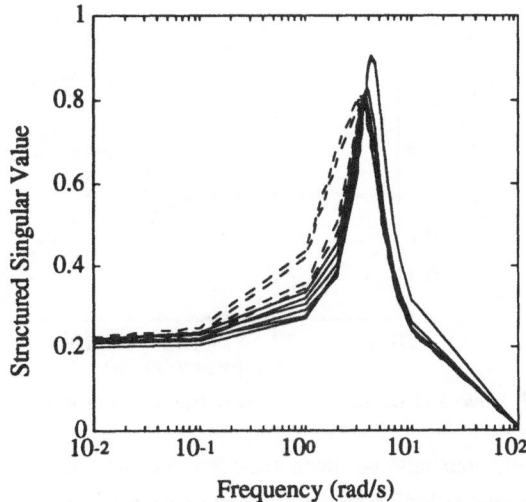

Fig. 4.23 Low-α μ Bounds for Robust Stability to Structured Uncertainty

For all conditions tested, μ is less than one for all frequencies. The desired level of stability robustness to structured uncertainty has been achieved.

4.7.1.2 Low Angle of Attack Unstructured Uncertainty

Unstructured uncertainty at the actuator input represents both unmodeled dynamics and saturation effects. Because structured singular value stability analysis methods have only recently been widely accepted, little data is available on what appropriate levels of uncertainty should be. The requirements for robustness for unstructured uncertainty for this design are derived from the classical gain margin specification of 6 db in each loop. This roughly corresponds to a 50% multiplicative uncertainty stability tolerance in individual loops. This requirement is extended to the case of simultaneous multiplicative perturbations by relaxing the level of uncertainty to 30%. Fig. 4.24 shows structured singular value analysis results for the design condition and test points 1-4 using

$$W_{act} = \begin{bmatrix} 0.3 & 0 & 0 \\ 0 & 0.3 & 0 \\ 0 & 0 & 0.3 \end{bmatrix} \tag{4.38}$$

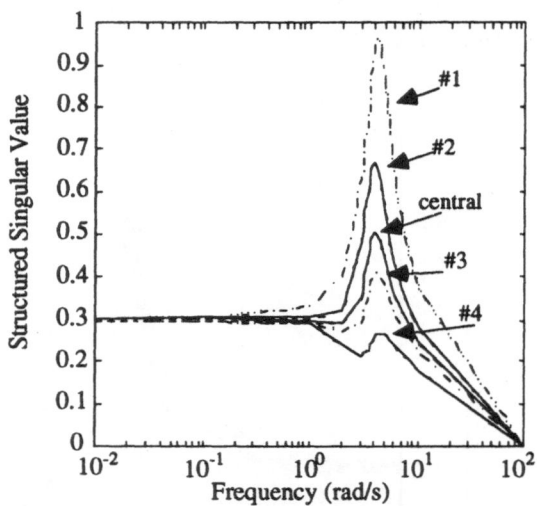

Fig. 4.24 Low-α μ Bounds for Unstructured Uncertainty at Plant Input

The desired level of robustness has been achieved to unstructured uncertainty at the plant input for all conditions tested. Analysis also reveals a trend, the lower dynamic pressure test cases exhibit lower levels of robustness for this type of uncertainty. The steady degradation of robustness is a result of higher feedback gains that are required for plant equalization at low dynamic pressure conditions.

Levels of unstructured uncertainty at the sensor output are driven by the quality of information available in each sensor channel. The body axis rotational rates can be measured by gyros which are precise and reliable. A sideslip signal, on the other hand, normally must be estimated or reconstructed using complementary filtering. Uncertainty in the sideslip measurement is the driving requirement for robustness analysis at the plant output. Structured singular value analysis is performed for a 10% multiplicative uncertainty in body axis rotational rate signals and a 40% multiplicative uncertainty in sideslip measurement.

$$W_{sens} = \begin{bmatrix} 0.4 & 0 & 0 \\ 0 & 0.1 & 0 \\ 0 & 0 & 0.1 \end{bmatrix} \tag{4.39}$$

The results for the design condition and test points 1-4 are shown in Fig. 4.25.

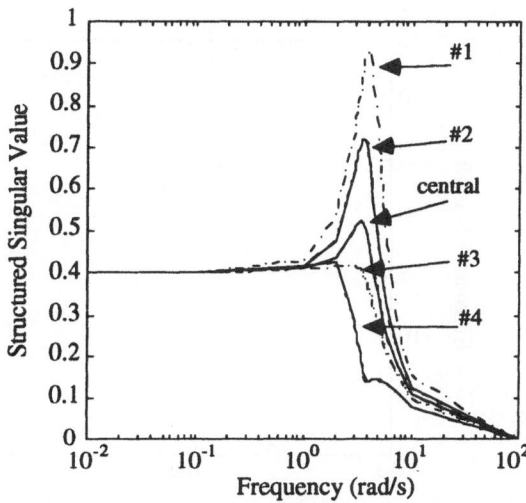

Fig. 4.25 Low-α μ Bounds for Unstructured Uncertainty at Plant Output

The desired level of robustness has been achieved to unstructured uncertainty at the sensor outputs for all conditions tested. Fig. 4.25 shows that the same reduction of robustness margins at low dynamic pressures is present in the output uncertainty analysis as it is with input uncertainty analysis.

4.7.1.3 Low Angle of Attack Robust Performance

Performance robustness is defined for this problem as achieving a small relative error between the *ideal model* and the actual response in the presence of uncertainties in aerodynamic parameters. The level of structured uncertainty included in this analysis is reduced from that used for robust stability analysis. The stability robustness analysis uncertainties were chosen to be worst case tests of closed loop stability. It is not reasonable to expect performance robustness to the same worst case set. Performance robustness analysis is performed using structured uncertainty levels of 25%, the size of those shown in Table 4.2. The performance weight, W_{perf}, used for this analysis is not the same as the weight used for control synthesis. That design weight was chosen with other considerations in mind, namely good nominal performance. The inverse of the performance weight, shown in Fig. 4.26, bounds the allowable error between the analysis model and the *ideal model* in the presence of structured uncertainties.

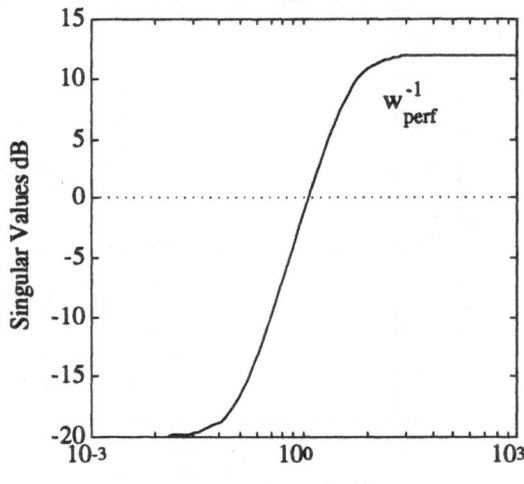

Fig. 4.26 Performance Error Bound for Robustness Analysis

The performance weight dictates that the steady state error to commands must be less than 10%. Fig. 4.27 shows the results of performance robustness analysis for the design point and test points 1-4.

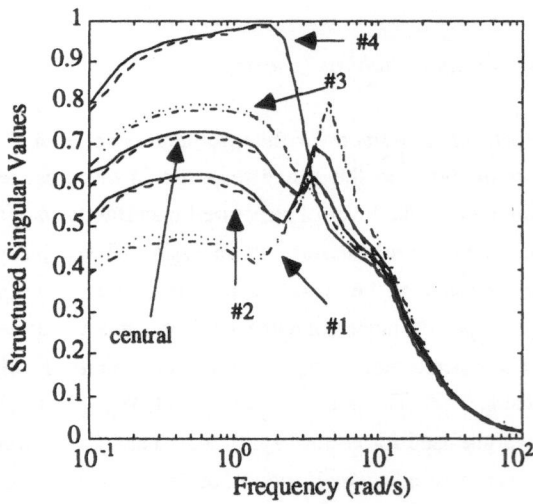

Fig. 4.27 Low-α μ Bounds for Performance Robustness

The structured singular values are less than unity for all frequencies, so performance robustness has been achieved for all conditions tested.

4.7.2 High Angle of Attack Robustness Results

4.7.2.1 High Angle of Attack Structured Uncertainty

Stability robustness to structured uncertainties is tested for the high-α *central* plant and the four test points described in Table 4.2. The levels of uncertainty in Table 4.3 are used for the analysis. Fig. 4.28 shows the resulting upper and lower bounds on the structured singular value.

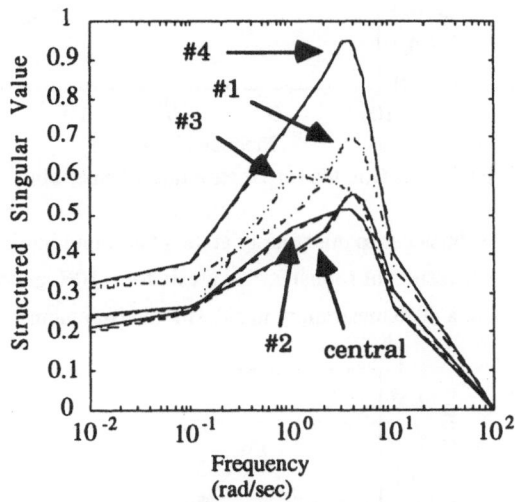

Fig. 4.28 High-α μ Bounds for Robust Stability to Structured Uncertainty

For all conditions tested μ is less than one for all frequencies. The desired level of stability robustness to structured uncertainty has been achieved.

4.7.2.2 High Angle of Attack Unstructured Uncertainty

Robustness to a 30% perturbation at the actuator input is tested. Fig. 4.29 shows structured singular value analysis results for the *central* plant and high-α test points 1-4.

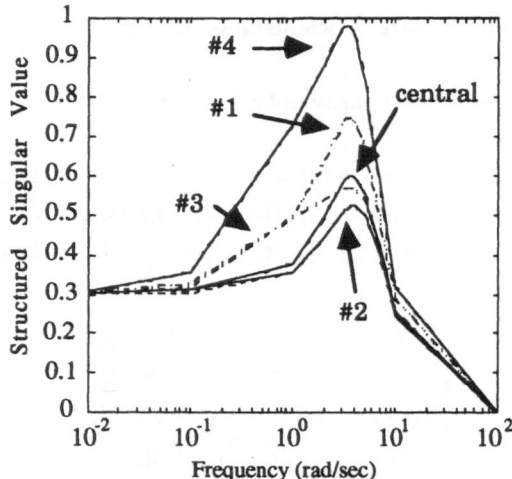

Fig. 4.29 High-α μ Bounds for Unstructured Uncertainty at Plant Input

The desired level of robustness to uncertainty at the plant input has been achieved.

Fig. 4.30 shows the results of robustness analysis for a 10% uncertainty in roll and yaw rate measurements and a 40% uncertainty in sideslip measurement.

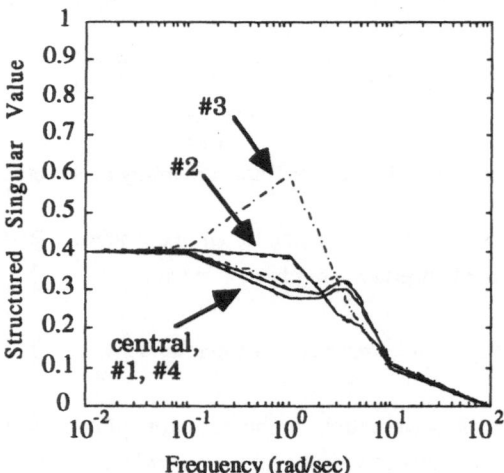

Fig. 4.30 High-α μ Bounds for Unstructured Uncertainty at Plant Output

For all conditions tested, the desired level of robustness to output uncertainties has been easily achieved for the *central* plant and the four test cases.

4.7.2.3 High Angle of Attack Robust Performance

Robust performance analysis is performed for the high-α *central* and test points using the same weights and levels of structured uncertainty described in section 4.7.1.3. Fig. 4.31 shows the results of this analysis.

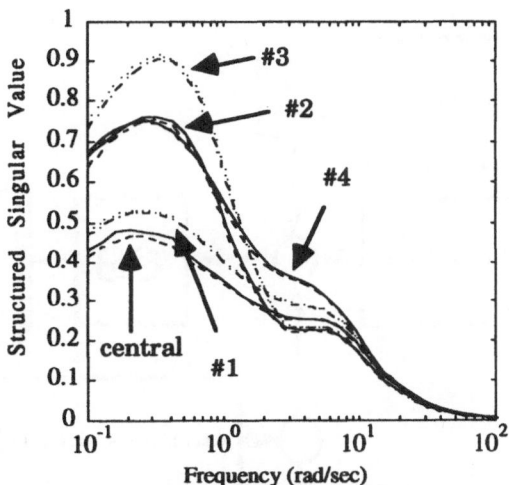

Fig. 4.31 High-α μ Bounds for Performance Robustness

The flight control system maintains good steady state tracking and satisfactory flying qualities in the face of uncertainties in aerodynamic and control derivatives.

4.7.3   Robustness Analysis of Low/High-α Blending

The controller blending introduced in section 4.6 creates a potential stability problem. While it has been proven that the low-α and high-α flight conditions are stabilized by their respective controllers, nothing has been demonstrated about the stability properties of the blending. Eqs. (4.28) and (4.29) can be rewritten such that the stability of the control blending becomes an uncertainty problem which can be solved with structured singular value methods. Redefine the blending parameter as

$$C = 0.5 \, (I + \Delta) \, , \quad \text{where} \quad \Delta = \delta I, \text{ and } \quad \bar{\sigma}(\Delta) \leq 1 \tag{4.40}$$

The desired dynamics and outer loop controller can now be represented as a function of the uncertain diagonal repeated block Δ.

$$v = 0.5(v_{\text{high-}\alpha} + v_{\text{low-}\alpha}) - 0.5\Delta(v_{\text{high-}\alpha} - v_{\text{low-}\alpha}) \qquad (4.41)$$

$$K_\mu = 0.5(K_{\mu_{\text{high-}\alpha}} + K_{\mu_{\text{low-}\alpha}}) - 0.5\Delta(K_{\mu\text{-high-}\alpha} - K_{\mu\text{-low-}\alpha}) \qquad (4.42)$$

Fig. 4.32 shows the robustness analysis model for control blending.

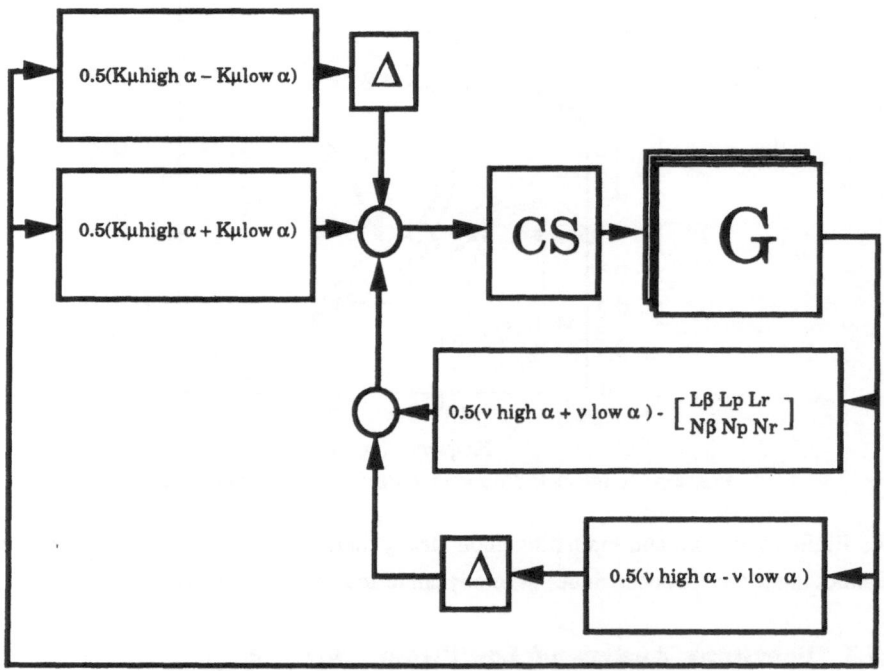

Fig. 4.32 Low/High-α Blending Robustness Model

If the system can tolerate a Δ with an infinity norm less than unity, the control blending is guaranteed not to destabilize the closed loop system for any value of C between 0 and 1. Fig. 4.33 shows the results of structured singular value analysis for the ten flight conditions described in Tables 4.1 and 4.2.

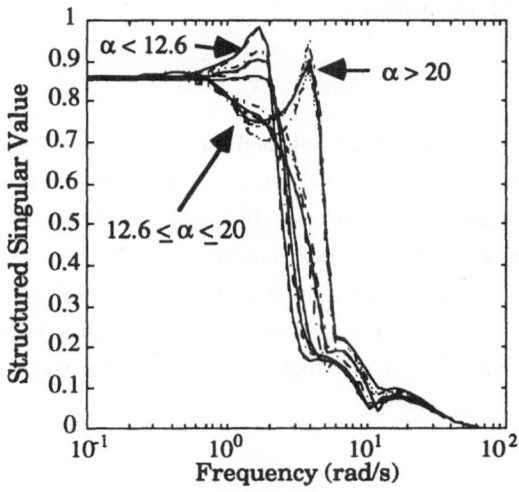

Fig. 4.33 Low/High-α Blending Robustness Results

The control blending is guaranteed not destabilize the system for any point tested. Notice that peaking occurs in the stuctured singular value plot for conditions where angle of attack is less than 12.6 degrees and greater than 20 degrees. In the region including angle of attack from 12.6 to 20 degrees, the structured singular values are very well behaved. The blending region described in section 4.6 should provide a smooth transition between the low- and high-α controllers.

4.8 Nonlinear Analysis

In order to be judged acceptable, a manual flight control system must not only meet the linear design criteria, it must exhibit satisfactory properties in a dynamic nonlinear simulation environment. A set of nonlinear evaluation manuevers is created to test the ability of the control system to handle large amplitude and rapidly changing inputs. These manuevers are performed at a range of flight conditions including both low and high angles of attack. Tracking performance and response decoupling are the primary measures of merit during these manuevers.

4.8.1   Conventional Maneuvers

(1) Loaded Roll

The aircraft is banked 60 deg to the right and loaded to 3-6 g's. While maintaining the g loading, a roll is executed at 60-120 deg/s to bring the aircraft to 60 deg left bank. The aircraft is then unloaded and the vehicle is rolled back to wings level. Fig. 4.34 shows a loaded roll maneuver. Fig. 4.38 shows the time histories for a loaded roll performed from a starting condition of Mach 0.7 and altitude 20,000 feet, an off design point. The time response plots show that tracking performance is maintained for large amplitude commands. Sideslip is regulated to less than 1 degree during the highly coupled maneuver.

(2) High Rate Roll

A large amplitude stability axis roll rate input is applied to roll the aircraft through 360 deg. Fig. 4.35 shows a high rate roll performed at Mach 0.6 and altitude 20,000 feet. The time response plots in Fig. 4.39 show that for a 250 deg/s stability axis roll rate command, good tracking and response decoupling are maintained. The maximum sideslip is less than 2.5 deg.

(3) Steady Sideslip

A 1 deg/s sideslip command is applied until a sideslip of 10 deg is achieved. The command is then reversed to bring the aircraft to 10 deg sideslip in the opposite direction. Fig. 4.36 shows a steady sideslip manuever performed at Mach 0.6 and altitude 20000 feet. The time responses in Fig. 4.40 show that the sideslip command is accurately tracked while a wings level attitude is maintained. Notice that at 10 deg sideslip, the rudder is close to position saturation.

(4) Turn Reversals

Large amplitude roll commands are applied to rapidly bring the aircraft to 60 deg left roll angle, then to 60 deg right roll angle, back to 60 deg left, and then to wings level. Fig. 4.37 shows a turn reversal manuever performed at Mach 0.6 and altitude 20,000 feet. The time responses in Fig. 4.41 show that the rapidly changing large amplitude stability axis roll rate commands test the ability of the control system to provide tracking and turn coordination while avoiding actuator saturations. The peak sideslip during this manuever is less than 3.0 deg.

Fig. 4.34 Loaded Roll Maneuver

Fig.4.35 High Rate Roll Maneuver

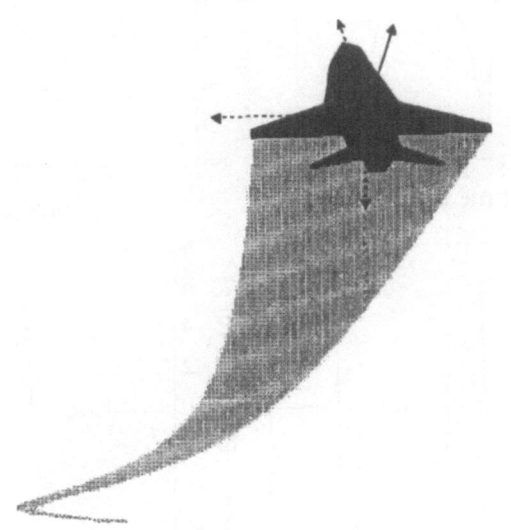

Fig. 4.36 Steady Sideslip Maneuver

Fig. 4.37 Turn Reversal Maneuver

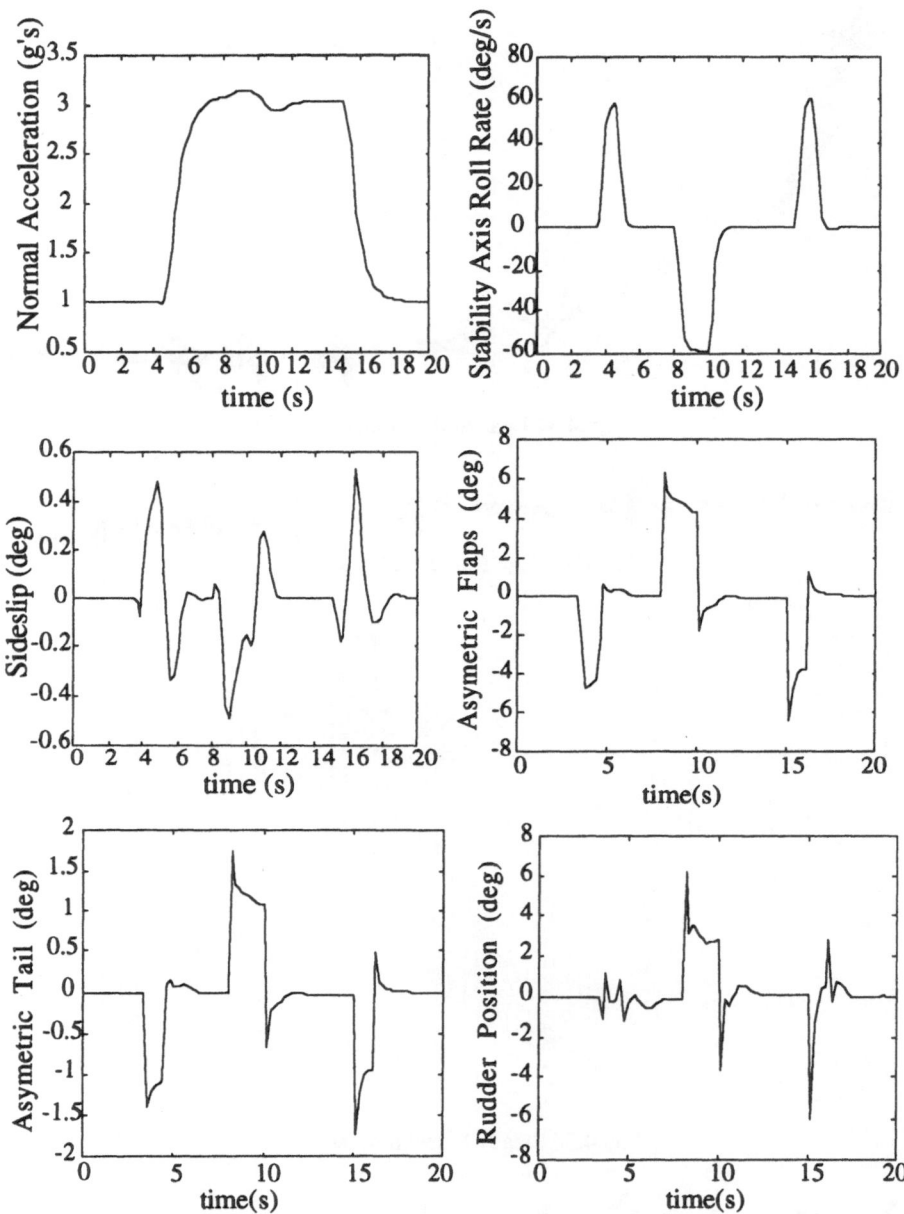

Fig. 4.38 Loaded Roll Maneuver Time Response

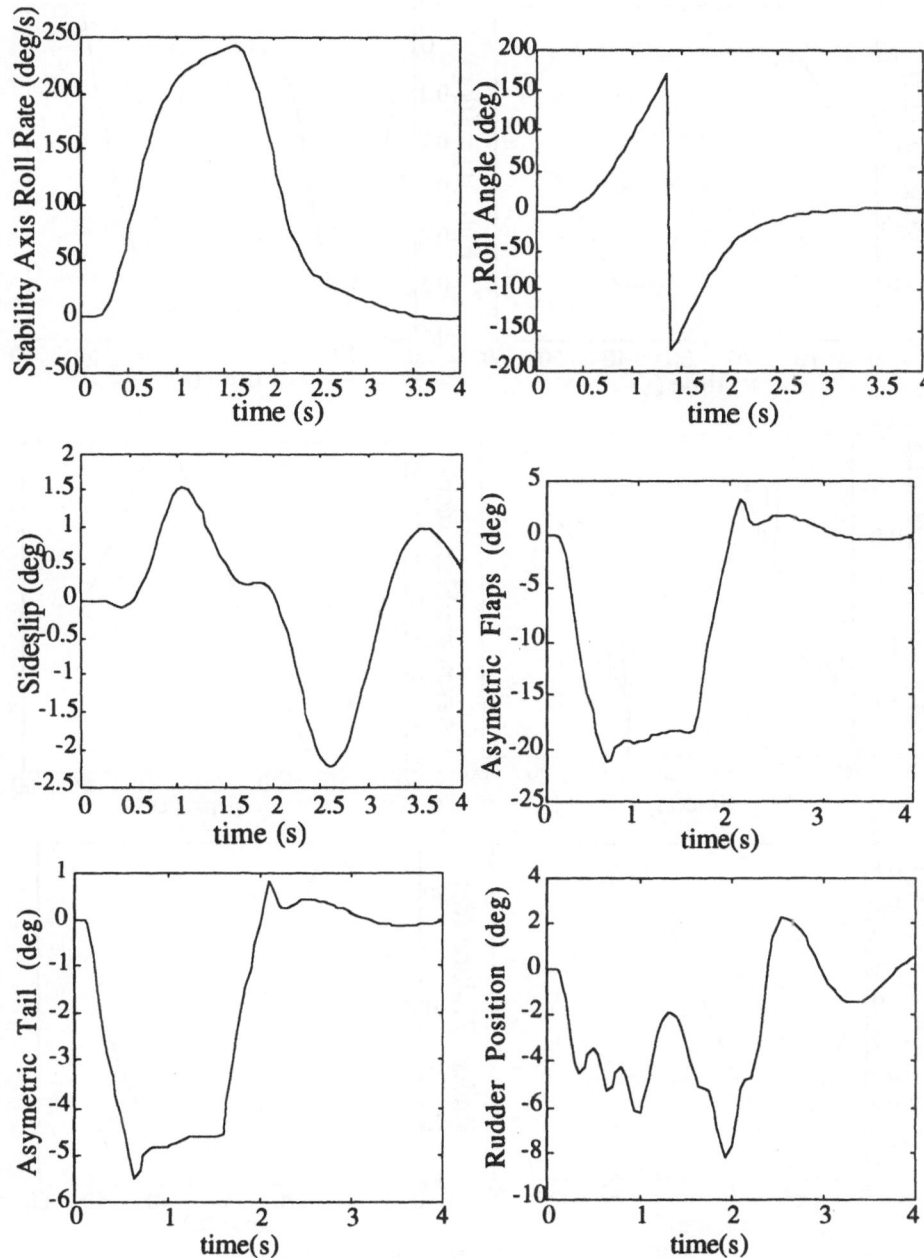

Fig. 4.39 High Rate Roll Maneuver Time Response

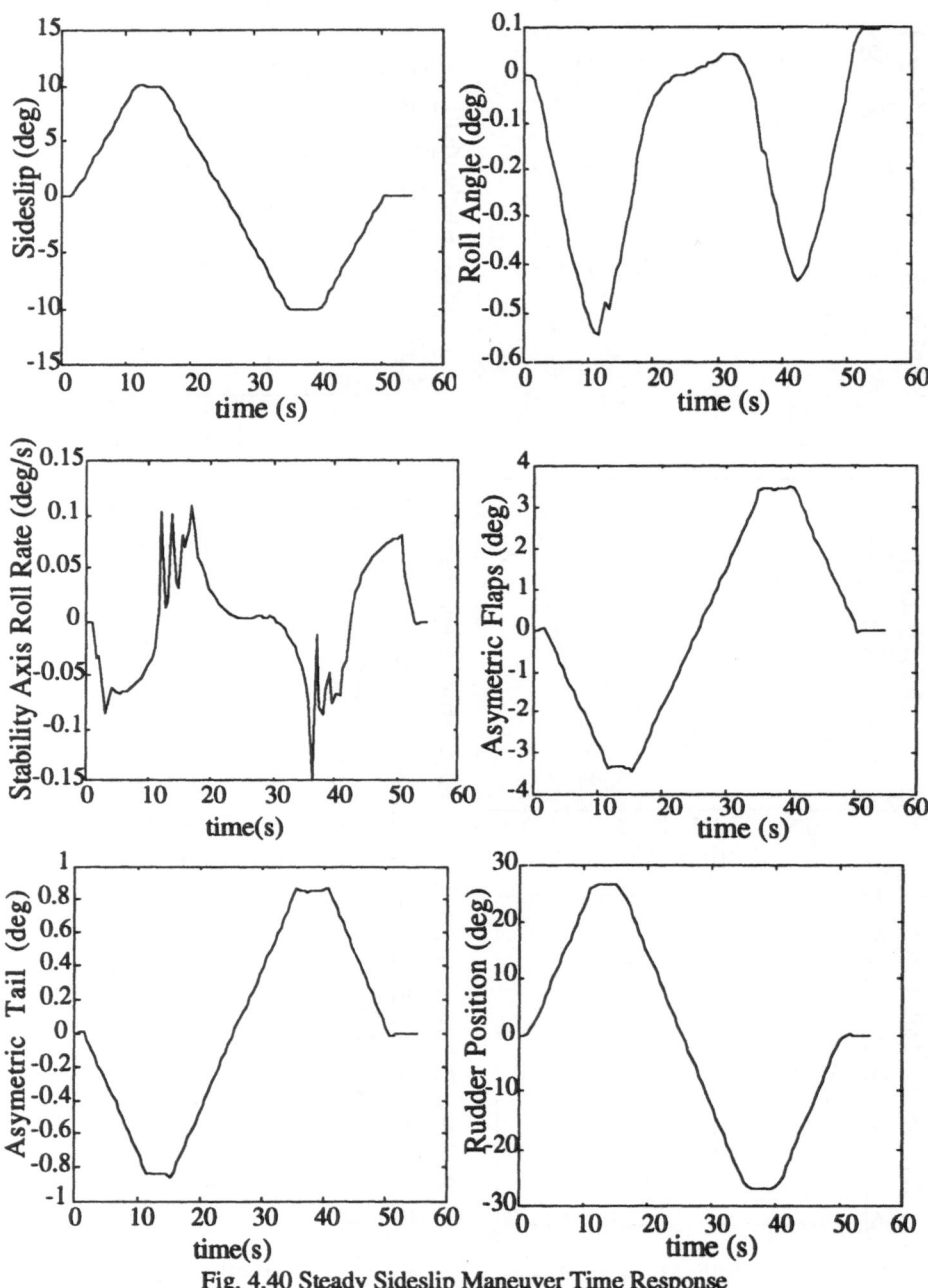

Fig. 4.40 Steady Sideslip Maneuver Time Response

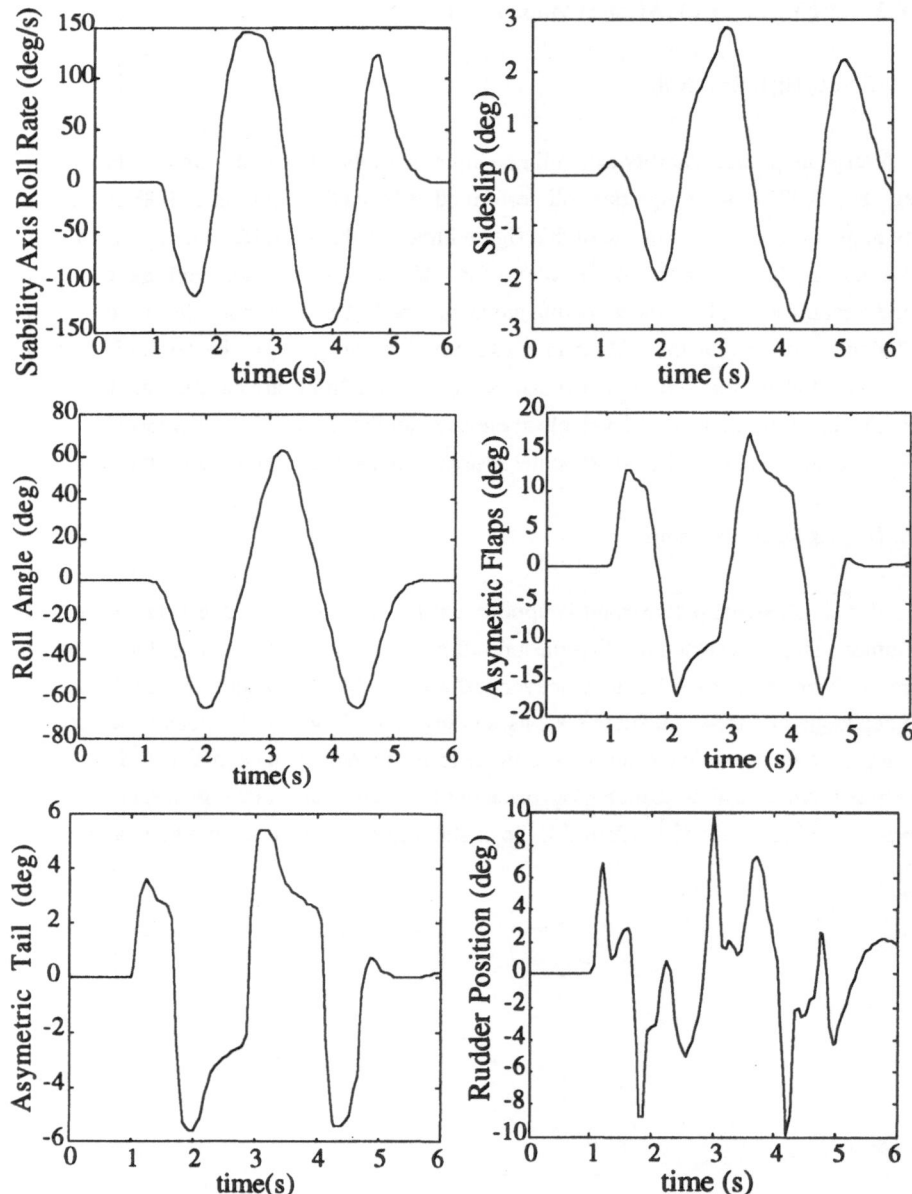

Fig. 4.41 Turn Reversal Maneuver Time Response

4.8.2   High Angle of Attack Maneuvers

(1) High-α High Rate Roll

A large amplitude stability axis roll rate input is applied to roll the aircraft through 360 deg. Fig. 4.42 shows a high rate roll performed at Mach 0.3, altitude 25,000 feet, and 23 deg angle of attack. The arrow in this figure indicates the aircraft's velocity vector. The time response plots in Fig. 4.44 show that for a 45 deg/s stability axis roll rate command, good tracking and response decoupling are maintained. The maximum sideslip is less than 4.1 deg. Notice that the rudder rate and position saturates at the beginning of this maneuver and asymmetric flaps also rate saturate. These limits are encountered because of the physical limitations of the vehicle at elevated angles of attack. It is important to note that the control system does not drive the system unstable when these saturations occur.

(2) High-α Steady Sideslip

A 1 deg/s sideslip command is applied until a sideslip of 10 deg is achieved. The command is then reversed to bring the aircraft back to 0 deg sideslip. Fig. 4.43 shows the aircraft during a steady sideslip maneuver performed at Mach 0.3, altitude 25,000 feet, and 23 deg angle of attack. The time responses in Fig. 4.45 show that the sideslip command is accurately tracked while a wings level attitude is maintained. Notice that at 10 deg sideslip, the maximum control deflection takes place in asymmetric flaps rather than rudder. This is because at high angles of attack, a sideslip or stability axis yaw requires a body axis roll.

Fig. 4.42 High-α High Rate Roll Maneuver

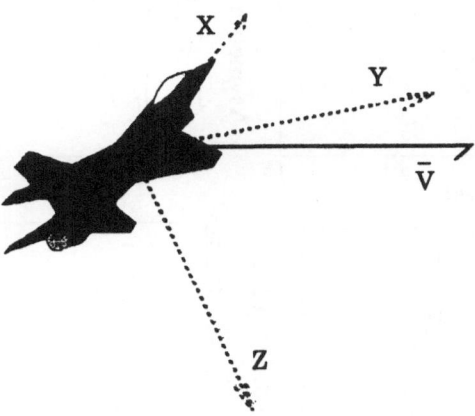

Fig. 4.43 High-α Steady Sideslip Maneuver

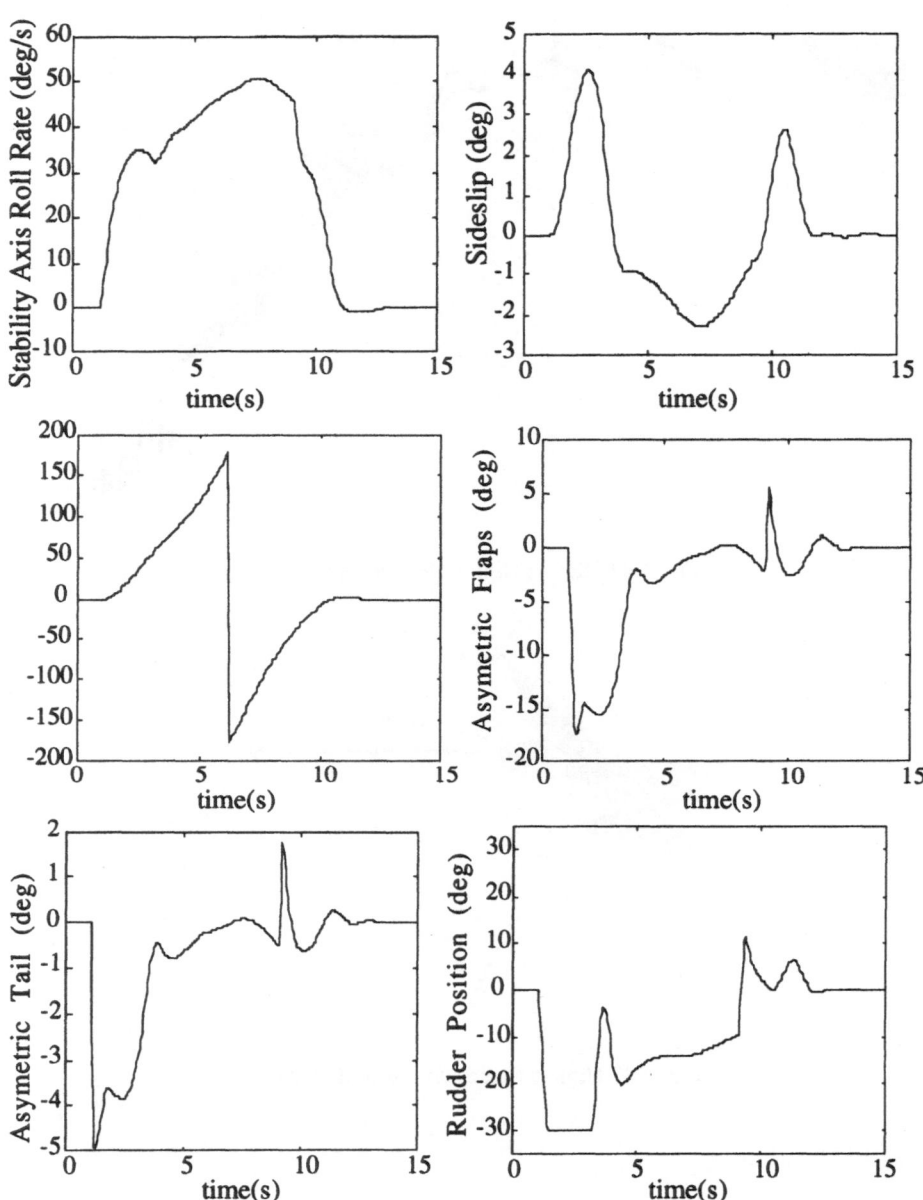

Fig. 4.44 High-α High Rate Roll Maneuver Time Response

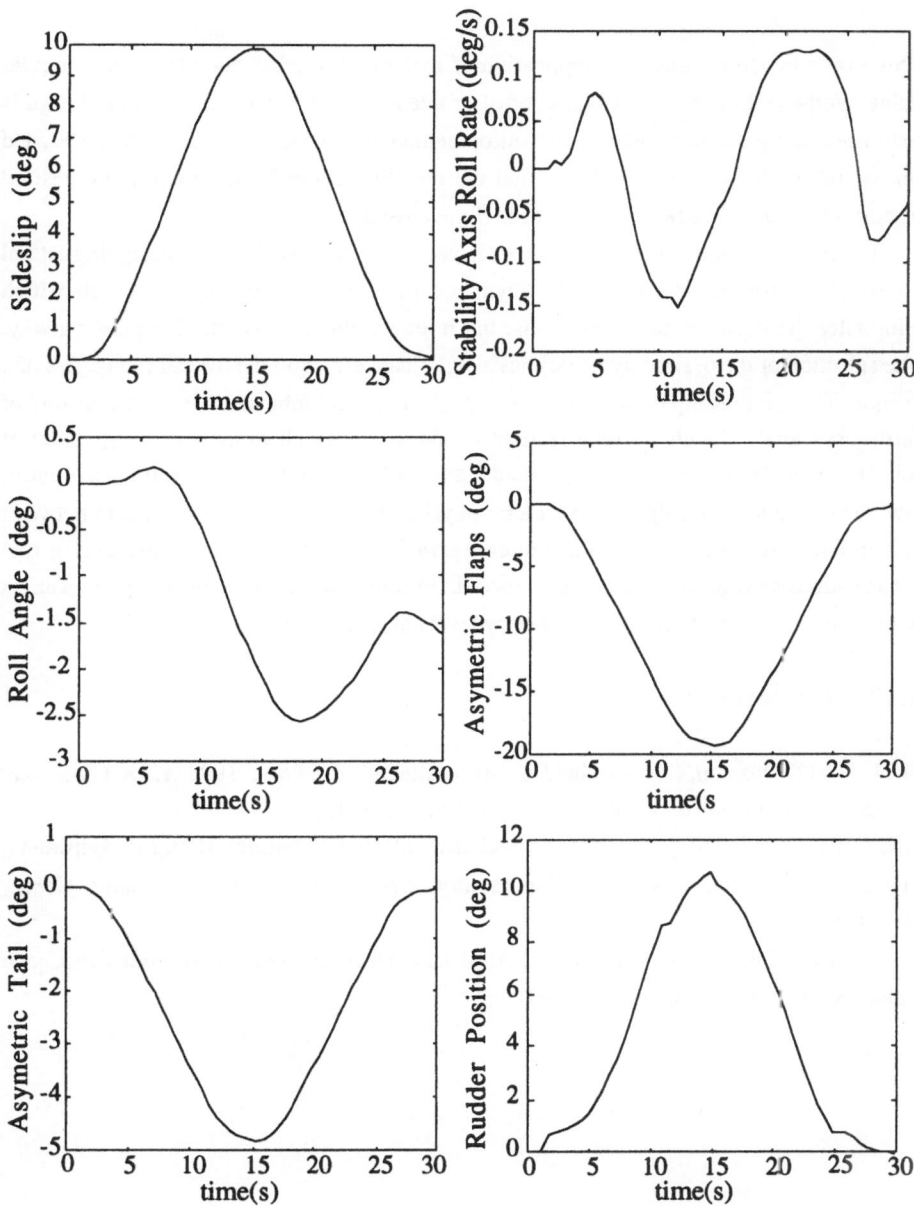

Fig. 4.45 High-α Steady Sideslip Maneuver Time Response

4.9   Conclusions and Lessons Learned

This example demonstrates an application of dynamic inversion and structured singular value synthesis to a manual flight control problem for the VISTA F-16. The design is performed using the inner/outer loop control methodology described in section 3. Detailed linear and nonlinear analysis shows that the resulting control laws provide the desired performance and robustness across a wide flight envelope.

A number of lessons learned can be derived from the experience of designing lateral control laws for the VISTA F-16. Dynamic inversion is a useful approach which eliminates the need for gain scheduling, but it may yield poor results if applied naively. The selection of the desired dynamics has a significant effect on the overall robustness of a design. The relationship between the desired dynamics and robustness is a topic worthy of further research. Another lesson learned is that it is difficult to include a high level of robustness in the μ-synthesis design model and still achieve a high performance control law. This is due primarily to two factors: physical limitations due to trade-offs between performance and robustness, and the conservatism that comes with representing real perturbations as complex in the design model. Fortunately, control laws designed using a reduced uncertainty set provide satisfactory levels of robustness.

4.10   References

[4.1] W. C. Durham, "Constrained Control Allocation," *Proc. 1992 AIAA Guidance, Navigation, and Control Conf.*, Hilton Head SC, Aug. 1992.

[4.2] J. C. Doyle, K. Lenz, and A. Packard, "Design Examples Using μ-Synthesis: Space Shuttle Lateral Axis FCS During Reentry," *Proc. 25th IEEE Conf. Decision Contr.*, Dec. 1986.

[4.3] K. R. Haiges et al., "Robust Control Law Development for Modern Aerospace Vehicles," WL-TR-91-3105, Aug. 1991.

Appendix 4

All of the linear models included in the appendix are given in the form:

$$
A = \begin{bmatrix} Y_\beta & \sin\alpha & -\cos\alpha \\ L_\beta & L_p & L_r \\ N_\beta & N_p & N_r \end{bmatrix}, \quad B = \begin{bmatrix} Y_{\delta T} & Y_{\delta DF} & Y_{\delta R} \\ L_{\delta T} & L_{\delta DF} & L_{\delta R} \\ N_{\delta T} & N_{\delta DF} & N_{\delta R} \end{bmatrix} \qquad \textbf{(A4.1)}
$$

LOW ANGLE OF ATTACK DESIGN AND ANALYSIS MATRICES

The state space representation of the low-α *central* model is:

$$
A_{central} = \begin{bmatrix} -1.6885e\text{-}01 & 7.5949e\text{-}02 & -9.9523e\text{-}01 \\ -2.7692e\text{+}01 & -2.3750e\text{+}00 & 1.7141e\text{-}01 \\ 6.6973e\text{+}00 & -6.6493e\text{-}02 & -3.9717e\text{-}01 \end{bmatrix} \qquad \textbf{(A4.2)}
$$

$$
B_{central} = \begin{bmatrix} 2.3384e\text{-}02 & 3.7619e\text{-}03 & 2.5281e\text{-}02 \\ -2.2464e\text{+}01 & -2.9507e\text{+}01 & 6.0951e\text{+}00 \\ -2.3973e\text{+}00 & -5.2764e\text{-}01 & -2.6622e\text{+}00 \end{bmatrix} \qquad \textbf{(A4.3)}
$$

The LQR gain matrix described in section 4.4.2 is:

$$
K_{LQ} = \begin{bmatrix} -8.3248e\text{-}01 & 1.7221e\text{-}01 & 1.0251e\text{-}01 \\ -2.6630e\text{+}00 & 1.1084e\text{-}01 & 2.0820e\text{+}00 \end{bmatrix} \qquad \textbf{(A4.4)}
$$

The low-α desired dynamics for the dynamic inversion calculations are :

$$
V_{low} = \begin{bmatrix} -2.6860e\text{+}01 & -2.5472e\text{+}00 & 6.8900e\text{-}02 \\ 9.3603e\text{+}00 & -1.7733e\text{-}01 & -2.4792e\text{+}00 \end{bmatrix} \begin{bmatrix} \beta \\ p \\ r \end{bmatrix} \qquad \textbf{(A4.5)}
$$

low-α test point #1 is:

$$
A = \begin{bmatrix} -8.8987e\text{-}02 & 2.1908e\text{-}01 & -9.7390e\text{-}01 \\ -1.6061e\text{+}01 & -1.0773e\text{+}00 & 3.2838e\text{-}01 \\ 2.1102e\text{+}00 & -8.5482e\text{-}02 & -2.2163e\text{-}01 \end{bmatrix} B = \begin{bmatrix} 1.1522e\text{-}02 & 3.1806e\text{-}04 & 1.3882e\text{-}02 \\ -7.0400e\text{+}00 & -1.0173e\text{+}01 & 2.3680e\text{+}00 \\ -7.5749e\text{-}01 & -9.0969e\text{-}02 & -9.3736e\text{-}01 \end{bmatrix} \textbf{(A4.6)}
$$

low-α test point #2 is:

$$A = \begin{bmatrix} -1.4115\text{e-}01 & 1.0824\text{e-}01 & -9.9214\text{e-}01 \\ -2.1711\text{e+}01 & -1.9630\text{e+}00 & 2.1997\text{e-}01 \\ 4.6980\text{e+}00 & -7.4018\text{e-}02 & -3.3259\text{e-}01 \end{bmatrix} \quad B = \begin{bmatrix} 1.8628\text{e-}02 & 1.7784\text{e-}03 & 2.0363\text{e-}02 \\ -1.5051\text{e+}01 & -2.0476\text{e+}01 & 4.4939\text{e+}00 \\ -1.7052\text{e+}00 & -3.1245\text{e-}01 & -1.8392\text{e+}00 \end{bmatrix} \quad \textbf{(A4.7)}$$

low-α test point #3 is:

$$A = \begin{bmatrix} -2.0556\text{e-}01 & 6.2737\text{e-}02 & -9.9570\text{e-}01 \\ -3.0469\text{e+}01 & -2.8680\text{e+}00 & 1.6159\text{e-}01 \\ 8.0291\text{e+}00 & -6.9031\text{e-}02 & -4.7919\text{e-}01 \end{bmatrix} \quad B = \begin{bmatrix} 2.7686\text{e-}02 & 5.2816\text{e-}03 & 3.0934\text{e-}02 \\ -2.7016\text{e+}01 & -3.5066\text{e+}01 & 7.3843\text{e+}00 \\ -2.9005\text{e+}00 & -6.6770\text{e-}01 & -3.2244\text{e+}00 \end{bmatrix} \quad \textbf{(A4.8)}$$

low-α test point #4 is:

$$A = \begin{bmatrix} -2.8053\text{e-}01 & 3.6571\text{e-}02 & -9.9663\text{e-}01 \\ -3.9901\text{e+}01 & -4.0119\text{e+}00 & 9.0623\text{e-}02 \\ 1.1859\text{e+}01 & -6.3609\text{e-}02 & -6.1759\text{e-}01 \end{bmatrix} \quad B = \begin{bmatrix} 3.5430\text{e-}02 & 7.3199\text{e-}03 & 4.4502\text{e-}02 \\ -3.7744\text{e+}01 & -4.9716\text{e+}01 & 1.1870\text{e+}01 \\ -4.4520\text{e+}00 & -1.0145\text{e+}00 & -5.4764\text{e+}00 \end{bmatrix} \quad \textbf{(A4.9)}$$

HIGH ANGLE OF ATTACK DESIGN AND ANALYSIS MATRICES

The state space representation of the high-α *central* model is:

$$A_{central} = \begin{bmatrix} -5.9262\text{e-}02 & 3.9572\text{e-}01 & -9.1655\text{e-}01 \\ -1.1812\text{e+}01 & -2.5948\text{e-}01 & 5.6269\text{e-}01 \\ 5.0798\text{e-}01 & -8.4359\text{e-}02 & -1.8858\text{e-}01 \end{bmatrix} \quad \textbf{(A4.10)}$$

$$B_{central} = \begin{bmatrix} 6.9396\text{e-}03 & -1.3504\text{e-}03 & 1.2277\text{e-}02 \\ -4.0377\text{e+}00 & -4.4641\text{e+}00 & 1.2137\text{e+}00 \\ -1.9196\text{e-}01 & 1.0752\text{e-}01 & -5.2916\text{e-}01 \end{bmatrix} \quad \textbf{(A4.11)}$$

The LQR gain matrix described in section 4.4.3 is:

$$K_{LQ} = \begin{bmatrix} -3.4574\text{e+}00 & 2.4418\text{e+}00 & 1.1895\text{e-}01 \\ -8.1359\text{e+}00 & 1.1895\text{e-}01 & 4.8183\text{e+}00 \end{bmatrix} \quad \textbf{(A4.12)}$$

The high-α desired dynamics for the dynamic inversion calculations are :

$$V_{high} = \begin{bmatrix} -8.3546\text{e+}00 & -2.7013\text{e+}00 & 4.4374\text{e-}01 \\ 8.6438\text{e+}00 & -2.0331\text{e-}01 & -5.0069\text{e+}00 \end{bmatrix} \begin{bmatrix} \beta \\ p \\ r \end{bmatrix} \quad \textbf{(A4.13)}$$

high-α test point #1 is:

$$A = \begin{bmatrix} -7.2864\text{e-}02 & 5.0042\text{e-}01 & -8.6388\text{e-}01 \\ -8.8073\text{e+}00 & -1.5020\text{e-}01 & 1.0335\text{e+}00 \\ -6.2957\text{e-}01 & -1.1955\text{e-}01 & -2.5384\text{e-}01 \end{bmatrix} \quad B = \begin{bmatrix} 9.6636\text{e-}03 & -1.7647\text{e-}03 & 1.6596\text{e-}02 \\ -3.9347\text{e+}00 & -2.7200\text{e+}00 & 1.0163\text{e+}00 \\ -1.1366\text{e-}01 & 1.4606\text{e-}01 & -4.4164\text{e-}01 \end{bmatrix} \quad \textbf{(A4.14)}$$

high-α test point #2 is:

$$A = \begin{bmatrix} -7.7420e\text{-}02 & 3.4247e\text{-}01 & -9.3725e\text{-}01 \\ -2.6269e\text{+}01 & -5.0772e\text{-}01 & 7.1424e\text{-}01 \\ 1.4662e\text{+}00 & -1.1973e\text{-}01 & -2.8648e\text{-}01 \end{bmatrix} B = \begin{bmatrix} 1.3011e\text{-}02 & -2.1880e\text{-}03 & 2.0275e\text{-}02 \\ -9.2229e\text{+}00 & -9.8389e\text{+}00 & 2.0911e\text{+}00 \\ -6.0950e\text{-}01 & 1.3102e\text{-}01 & -1.1832e\text{+}00 \end{bmatrix} (A4.15)$$

high-α test point #3 is:

$$A = \begin{bmatrix} -1.0581e\text{-}01 & 3.4259e\text{-}01 & -9.3671e\text{-}01 \\ -3.7254e\text{+}01 & -7.6492e\text{-}01 & 1.0747e\text{+}00 \\ 9.9705e\text{-}01 & -1.8026e\text{-}01 & -4.3128e\text{-}01 \end{bmatrix} B = \begin{bmatrix} 1.9012e\text{-}02 & -4.1297e\text{-}03 & 2.9336e\text{-}02 \\ -1.7573e\text{+}01 & -1.6894e\text{+}01 & 3.7761e\text{+}00 \\ -1.1659e\text{+}00 & 2.5008e\text{-}01 & -2.2423e\text{+}00 \end{bmatrix} (A4.16)$$

high-α test point #4 is:

$$A = \begin{bmatrix} -3.6933e\text{-}02 & 5.0019e\text{-}01 & -8.6457e\text{-}01 \\ -1.2679e\text{+}01 & -1.7145e\text{-}01 & 1.1820e\text{+}00 \\ 2.0841e\text{+}00 & -1.3696e\text{-}01 & -2.9045e\text{-}01 \end{bmatrix} B = \begin{bmatrix} 9.0561e\text{-}03 & -3.1149e\text{-}03 & 1.5815e\text{-}02 \\ -7.9796e\text{+}00 & -5.4067e\text{+}00 & 2.2032e\text{+}00 \\ -1.7620e\text{-}01 & 3.1292e\text{-}01 & -9.5343e\text{-}01 \end{bmatrix} (A4.17)$$

OUTER LOOP CONTROLLER MATRICES

The full and reduced order outer loop controllers have the following form:

$$K_\mu = \frac{\begin{bmatrix} num11 & num12 \\ num21 & num22 \end{bmatrix}}{den} , \qquad\qquad (A4.18)$$

where

$$num = K(s + z_1)(s + z_2)...(s + z_{n-1})(s + z_n) , \qquad\qquad (A4.19)$$

and

$$den = (s + p_1)(s + p_2)...(s + p_{n-1})(s + p_n) . \qquad\qquad (A4.20)$$

Table A4.1 Full Order Low-α Outer Loop Controller

	num11	num12	num21	num22	den
K (p_{22})	-2.5721e+03	-8.3453e+01	4.6524e+01	4.0843e+03	-6.9474e+02
z_{21} (p_{21})	-6.9474e+02	-6.9443e+02	-6.9477e+02	-6.9444e+02	-6.9469e+02
z_{20} (p_{20})	-6.9469e+02	-6.9469e+02	-6.9468e+02	-6.9469e+02	-6.9469e+02
z_{19} (p_{19})	-6.9469e+02	-6.9469e+02	-6.9469e+02	-6.9469e+02	-6.9469e+02
z_{18} (p_{18})	-9.0651e+01	-8.9410e+01	-8.8507e+01	-4.6643e+01+ 4.5350e+01i	-9.0650e+01
z_{17} (p_{17})	-4.7124e+01+ 4.5254e+01i	-4.9275e+01+ 4.5883e+01i	-4.3328e+01 +4.5444e+01i	-4.6643e+01- 4.5350e+01i	-8.9090e+01
z_{16} (p_{16})	-4.7124e+01- 4.5254e+01i	-4.9275e+01- 4.5883e+01i	-4.3328e+01- 4.5444e+01i	-8.9090e+01	-2.0971e+01+ 2.5276e+01i
z_{15} (p_{15})	-2.0982e+01+ 2.5265e+01i	6.5960e-01+ 2.6938e+01i	-1.0606e+00 +4.4836e+01i	-1.9293e+01	-2.0971e+01- 2.5276e+01i
z_{14} (p_{14})	-2.0982e+01- 2.5265e+01i	6.5960e-01- 2.6938e+01i	-1.0606e+00- 4.4836e+01i	-5.5413e+00+ 1.0974e+01i	-5.4138e+00+ 1.0886e+01i
z_{13} (p_{13})	-1.6045e+01	-1.5847e+01+ 1.7131e+00i	-3.2032e+01	-5.5413e+00- 1.0974e+01i	-5.4138e+00- 1.0886e+01i
z_{12} (p_{12})	-1.6385e+01	-1.5847e+01- 1.7131e+00i	-1.8881e+01 +1.2690e+01i	-1.2267e+01+ 2.0950e+00i	-1.6021e+01
z_{11} (p_{11})	-5.3011e+00+ 4.2606e+00i	1.6723e+00+ 4.9997e+00i	-1.8881e+01- 1.2690e+01i	-1.2267e+01- 2.0950e+00i	-1.1339e+01
z_{10} (p_{10})	-5.3011e+00- 4.2606e+00i	1.6723e+00- 4.9997e+00i	-1.6102e+01	-5.5771e+00+ 1.2845e+00i	-9.7888e+00
z_9 (p_9)	-1.4013e+00+ 3.4410e+00i	-3.8339e+00+ 1.4163e+00i	-5.3073e+00 +4.2694e+00i	-5.5771e+00- 1.2845e+00i	-4.7406e+00
z_8 (p_8)	-1.4013e+00- 3.4410e+00i	-3.8339e+00- 1.4163e+00i	-5.3073e+00- 4.2694e+00i	-2.6205e+00	-1.8338e+00
z_7 (p_7)	-1.8380e+00	-1.8380e+00	-1.8380e+00	-1.8380e+00	-1.8380e+00
z_6 (p_6)	-1.8379e+00	-1.8292e+00	-1.8380e+00	-1.8365e+00	-8.2088e-04+ 1.8614e-04i
z_5 (p_5)	-7.3952e-04	1.1408e-02	-7.0672e-04	-2.3573e-03	-8.2088e-04- 1.8614e-04i
z_4 (p_4)	-3.5229e-02+ 1.5362e-07i	-7.6371e-03	-7.1055e-03	-7.1499e-03	-3.5266e-02
z_3 (p_3)	-3.5229e-02- 1.5362e-07i	-3.5464e-02	-3.5228e-02	-7.1055e-03	-3.5229e-02
z_2 (p_2)	-7.1054e-03	-7.1055e-03	-3.5229e-02	-3.5236e-02	-7.1055e-03
z_1 (p_1)	-7.1055e-03	-3.5229e-02	-7.1061e-03	-3.5229e-02	-7.1055e-03

Table A4.2 Full Order High-α Outer Loop Controller

	num11	num12	num21	num22	den
K (p_{26})	1.4266e+02	-9.6735e+00	-3.5741e+00	3.4750e+02	-4.7660e+02
z_{25} (p_{25})	-4.7660e+02	-4.7656e+02	-4.7660e+02	-4.7655e+02	-1.0012e+02
z_{24} (p_{24})	-4.7659e+02	-4.7659e+02	-4.7659e+02	-4.7659e+02	-9.9999e+01
z_{23} (p_{23})	-4.7659e+02	-4.7659e+02	-4.7659e+02	-4.7659e+02	-4.7659e+02
z_{22} (p_{22})	-1.0012e+02	-1.0001e+02	-1.0002e+02	-4.7346e+01+4.5237e+01i	-4.7659e+02+1.6640e-10i
z_{21} (p_{21})	-4.7699e+01+4.5164e+01i	-5.3824e+01+4.6702e+01i	-5.2048e+01+4.6623e+01i	-4.7346e+01-4.5237e+01i	-4.7659e+02-1.6640e-10i
z_{20} (p_{20})	-4.7699e+01-4.5164e+01i	-5.3824e+01-4.6702e+01i	-5.2048e+01-4.6623e+01i	-9.9999e+01	-1.1082e+01+1.5128e+01i
z_{19} (p_{19})	-1.1096e+01+1.5150e+01i	4.1844e+00+2.7672e+01i	3.8206e+01	-2.3726e+01	-1.1082e+01-1.5128e+01i
z_{18} (p_{18})	-1.1096e+01-1.5150e+01i	4.1844e+00-2.7672e+01i	-2.3858e+01	-1.4165e+01	-2.3724e+01
z_{17} (p_{17})	-2.3745e+01	-2.3731e+01	-1.1245e+01+1.2516e+01i	-3.6793e+00+7.7239e+00i	-1.5414e+01
z_{16} (p_{16})	-1.5330e+01	-9.8520e+00+2.5888e+00i	-1.1245e+01-1.2516e+01i	-3.6793e+00-7.7239e+00i	-3.6829e+00+7.7075e+00i
z_{15} (p_{15})	-1.1196e+01	-9.8520e+00-2.5888e+00i	-1.5306e+01	-9.0016e+00+2.9329e+00i	-3.6829e+00-7.7075e+00i
z_{14} (p_{14})	-3.3301e+00+2.9883e+00i	-3.2448e+00+6.5110e+00i	-2.9927e+00+9.2275e+00i	-9.0016e+00-2.9329e+00i	-8.1682e+00+2.4218e+00i
z_{13} (p_{13})	-3.3301e+00-2.9883e+00i	-3.2448e+00-6.5110e+00i	-2.9927e+00-9.2275e+00i	-5.0965e+00	-8.1682e+00-2.4218e+00i
z_{12} (p_{12})	-3.5634e+00+2.9423e+00i	-5.9327e+00	-3.5404e+00+2.9777e+00i	-4.6048e+00	-2.0084e+00
z_{11} (p_{11})	-3.5634e+00-2.9423e+00i	-2.0469e+00	-3.5404e+00-2.9777e+00i	-1.9867e+00	-3.0581e+00+5.4067e-03i
z_{10} (p_{10})	-3.0396e+00	-2.9859e+00	-3.0368e+00	-3.2284e+00	-3.0581e+00-5.4067e-03i
z_9 (p_9)	-3.0520e+00+1.2596e-05i	-3.0520e+00	-3.0520e+00	-3.0520e+00	-3.0520e+00
z_8 (p_8)	-3.0520e+00-1.2596e-05i	-3.0515e+00	-3.0520e+00	-3.0525e+00	-1.3347e-02
z_7 (p_7)	-2.0863e-04	-1.3502e-02	-1.3348e-02	-1.3376e-02	-2.5157e-03
z_6 (p_6)	-1.3347e-02	2.0155e-04	-1.9525e-04	-1.1196e-04	-7.6053e-05
z_5 (p_5)	-2.5228e-03	-4.7022e-03	-2.5139e-03	-4.6981e-03	-2.0341e-04
z_4 (p_4)	-2.5357e-02	-4.6725e-03	-4.6745e-03	-4.6725e-03	-2.5356e-02
z_3 (p_3)	-2.5359e-02	-2.6712e-03	-4.6725e-03	-2.5357e-02	-2.5357e-02
z_2 (p_2)	-4.6739e-03	-2.5357e-02	-2.5357e-02	-2.5356e-02	-4.6725e-03
z_1 (p_1)	-4.6725e-03	-2.5356e-02	-2.5356e-02	-2.5216e-03	-4.6725e-03

Table A4.3 Reduced Order Low-α Outer Loop Controller

	num11	num12	num21	num22	den
K	-9.1780e-03	-1.3913e-04	-5.2125e-04	-7.9018e-06	
z_{15} (p_{15})	2.8279e+05	-5.9748e+05	9.1881e+04	5.1692e+08	-7.0782e+02
z_{14} (p_{14})	-6.9539e+02	-7.1262e+02	-6.6096e+02	-7.0782e+02	-6.9537e+02
z_{13} (p_{13})	-8.7801e+01	-8.1199e+01	-1.2255e+02	-4.6590e+01+ 4.6480e+01i	-8.7837e+01
z_{12} (p_{12})	-4.5073e+01+ 4.5590e+01i	-4.8789e+01+ 4.6977e+01i	-3.7043e+01 +5.4802e+01i	-4.6590e+01- 4.6480e+01i	-8.1139e+01
z_{11} (p_{11})	-4.5073e+01- 4.5590e+01i	-4.8789e+01- 4.6977e+01i	-3.7043e+01- 5.4802e+01i	-8.1139e+01	-2.1317e+01+ 2.4641e+01i
z_{10} (p_{10})	-2.1323e+01+ 2.4620e+01i	6.7676e-01+ 2.6868e+01i	-8.9503e-01 +4.4605e+01i	-5.4933e+00+ 1.1069e+01i	-2.1317e+01- 2.4641e+01i
z_9 (p_9)	-2.1323e+01- 2.4620e+01i	6.7676e-01- 2.6868e+01i	-8.9503e-01- 4.4605e+01i	-5.4933e+00- 1.1069e+01i	-5.3710e+00+ 1.0982e+01i
z_8 (p_8)	-5.0555e+00+ 3.6437e+00i	1.7480e+00+ 5.0408e+00i	-1.7632e+01+ 1.7084e+01i	-1.2468e+01	-5.3710e+00- 1.0982e+01i
z_7 (p_7)	-5.0555e+00- 3.6437e+00i	1.7480e+00- 5.0408e+00i	-1.7632e+01- 1.7084e+01i	-4.6269e+00+ 1.7637e+00i	-5.2005e+00+ 9.0751e-01i
z_6 (p_6)	-1.4084e+00+ 3.4439e+00i	-3.3998e+00+ 1.2372e+00i	-5.2133e+00 +3.7099e+00i	-4.6269e+00- 1.7637e+00i	-5.2005e+00- 9.0751e-01i
z_5 (p_5)	-1.4084e+00- 3.4439e+00i	-3.3998e+00- 1.2372e+00i	-5.2133e+00- 3.7099e+00i	-2.5910e+00	-3.3248e-02
z_4 (p_4)	-3.3212e-02	1.1408e-02	-3.3213e-02	-8.9760e-03	-8.2088e-04+ 1.8614e-04i
z_3 (p_3)	-8.9757e-03	-8.9639e-03	-8.9752e-03	-2.3573e-03	-8.2088e-04- 1.8614e-04i
z_2 (p_2)	-7.3952e-04	-3.3448e-02	-7.1275e-03	-7.1710e-03	-7.1273e-03
z_1 (p_1)	-7.1264e-03	-7.6699e-03	-7.0672e-04	-3.3218e-02	-8.9744e-03

Table A4.4 Reduced Order High-α Outer Loop Controller

	num11	num12	num21	num22	den
K	-2.0973e-04	4.4100e-03	2.7512e-03	-5.7849e-02	
z_{12} (p_{12})	9.0175e+05	3.7884e+03	2.4920e+03	8.1013e+03	-9.3515e+02
z_{11} (p_{11})	-8.2919e+02	-6.9228e+02	-6.7712e+02	-9.3391e+02	-8.2879e+02
z_{10} (p_{10})	-5.5991e+01	-5.4575e+01	-5.0619e+01	-2.0911e+01+ 1.1223e+01i	-7.9852e+00+ 1.7716e+01i
z_9 (p_9)	-8.0142e+00+ 1.7728e+01i	5.4051e+00+ 2.6664e+01i	3.6931e+01	-2.0911e+01- 1.1223e+01i	-7.9852e+00- 1.7716e+01i
z_8 (p_8)	-8.0142e+00- 1.7728e+01i	5.4051e+00- 2.6664e+01i	-5.3105e+00+ 1.4401e+01i	-3.9823e+00+ 8.1873e+00i	-3.9741e+00+ 8.1974e+00i
z_7 (p_7)	-6.3423e+00	-3.1426e+00+ 7.0307e+00i	-5.3105e+00- 1.4401e+01i	-3.9823e+00- 8.1873e+00i	-3.9741e+00- 8.1974e+00i
z_6 (p_6)	-2.3853e+00+ 2.9159e+00i	-3.1426e+00- 7.0307e+00i	-4.1728e+00+ 6.0155e+00i	-2.4193e+00+ 6.6166e-01i	-2.1405e+00
z_5 (p_5)	-2.3853e+00- 2.9159e+00i	-1.6954e+00	-4.1728e+00- 6.0155e+00i	-2.4193e+00- 6.6166e-01i	-2.6792e-02
z_4 (p_4)	-2.6804e-02	-2.7193e-02	-2.5690e-03	-2.6817e-02	-5.8955e-03
z_3 (p_3)	-2.5794e-03	-2.7396e-03	-5.9017e-03	-2.5795e-03	-2.0341e-04
z_2 (p_2)	-5.8947e-03	-5.9283e-03	-2.6786e-02	-5.9316e-03	-7.6053e-05
z_1 (p_1)	-2.0863e-04	2.0155e-04	-1.9525e-04	-1.1195e-04	-2.5715e-03

CHAPTER 5

THRUST VECTORING F-18 DESIGN

This section presents the design of a full envelope flight control system for a modified F-18 aircraft. The purpose of the design is to take advantage of advanced linear control design techniques while addressing practical implementation issues. A description of the aircraft model is presented, followed by detailed descriptions of the flight control system design and analysis. The specific numbers used in the design are indicated in Appendix 5.

5.1 Model Description

The aircraft model described in this section is based upon a modified version of the F-18 aircraft. The F-18 aircraft is a twin engine fighter aircraft with a moderately swept wing, twin canted vertical tails, and a large leading edge root extension. Leading and trailing edge flaps are used to control the wing camber for maximum maneuvering performance throughout the flight envelope. The horizontal tail is located below the wing to provide increased longitudinal stability at high angles of attack. The twin vertical tails provide positive directional stability beyond maximum trimmed angles of attack. An F-18 aircraft is shown in Fig. 5.1.

The aircraft model is augmented with two-dimensional thrust vectoring nozzles that provide pitch and yaw moments when deflected symmetrically and a roll moment when deflected asymmetrically. The aerodynamic control inputs to the aircraft dynamics are the elevators, the ailerons, the rudders, and the leading and trailing edge flaps. The aerodynamic surfaces are useful at normal flight conditions, where there is adequate aerodynamic control surface effectiveness. The thrust vectoring inputs are useful at high angle of attack, low dynamic pressure operating conditions, where the traditional aerodynamic control effectiveness is inadequate. The pilot inputs include a control stick and rudder pedals.

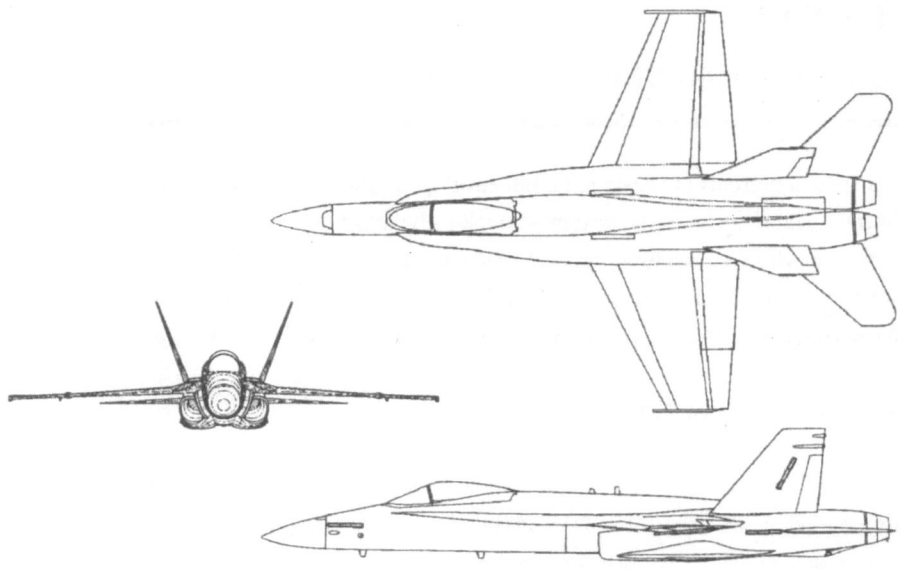

Fig. 5.1 F-18 Aircraft

5.1.1 Nonlinear Model

The aircraft model is a nonlinear FORTRAN model. The model consists of separate modules describing the atmosphere, nonlinear equations of motion, aerodynamics, engines including thrust vectoring nozzles, variable geometry inlets, sensors, and actuators which include rate and position limits. The high-fidelity model was developed in [5.1], which gives more detail than presented here. This section is presented for completeness; however, it is only a summary of the model description given in [5.1]. The nonlinear model structure of the thrust vectoring F-18 aircraft is the same as the F-16 model structure shown in Fig. 4.2.

The central component of the nonlinear model contains the six-degree-of-freedom aircraft equations of motion given by eqs. (2.1) and (2.2). A complete description of the equations of motion and atmospheric components is given in section 2.1. The primary inputs to the equations of motion are the forces and moments generated by the aerodynamic, propulsion, and inlet systems. The primary outputs are the state variables of the aircraft.

The aerodynamic actuator component of the nonlinear model transforms actuator commands into aerodynamic surface positions. There are five pairs of aerodynamic surfaces: three pairs for active control and two pair scheduled for optimum performance. The ailerons, rudders, and elevators are used for stability augmentation and flight path manipulation. The leading and trailing edge flaps are scheduled versus angle of attack, Mach number, and dynamic pressure to provide optimal lift-to-drag ratio, improve high angle of attack stability, and alleviate air loads on the wing during high-g maneuvers. All aerodynamic surfaces are powered by hydro-mechanical servo-actuators. The leading edge flaps use rotary hydraulic drive motors to power mechanical actuators while all other surfaces use linear piston actuators. The dynamic models of the aerodynamic surface actuators and corresponding position and rate limits are given in Appendix 5. The resulting aerodynamic surface positions from the actuator models are input to the aerodynamic model.

Baseline aerodynamic data are obtained from wind tunnel and flight test data covering Mach numbers ranging from 0.2 to 2, altitudes from sea level to 60,000 feet, and angles of attack and sideslip ranging from -12 to 90 degrees and -20 to 20 degrees respectively. Thrust vectoring-induced aerodynamic data are obtained from nonaxisymmetric nozzle wind tunnel tests and predictive jet interference analytical methods. Linear extrapolation is used to define aircraft properties outside of the ranges of the wind tunnel/flight test model. The aerodynamic data are contained in tabular format and linear interpolation is used for traditional force and moment aerodynamic coefficient build-up. Thrust vectoring-induced aerodynamic effects contributions are added to static and dynamic baseline aerodynamic coefficients to obtain total aerodynamic coefficients. The total coefficients are translated from the stability axes to the body axes and modified to account for center of gravity offset.

A modified engine model is used to provide increased thrust to the modified F-18 airframe model, and the airflow is scaled to match standard F404-400 engine airflow. The model represents the controlled closed-loop response of the nozzle and engine. Engine dynamics are modeled as a first order lag with a time constant of 0.8 second and rate limits of ±22 deg/sec to model engine control acceleration/deceleration schedules.

Symmetric and asymmetric motion of axisymmetric gimbaled nozzles provide roll, pitch, and yaw moments by vectoring the engine exhaust. Over/underturn effects of the vectored flow are modeled, and the gimbaled nozzle actuators are assumed to have second order dynamics with position and rate limits. The nozzle actuator models are given in Appendix 5. The inputs to the propulsion model are thrust commands, flight conditions, and thrust vectoring commands, and the outputs are propulsion forces and moments. Thrust and thrust vectoring forces and moments are transformed into the body axes and center of gravity offset is taken into account.

The F-18 model is modified with a high performance variable geometry inlet model. The inlet model consists of a fixed ramp and an adjustable second ramp to provide external supersonic flow compression at off-design conditions. Auxiliary by-pass and inlet doors in the subsonic diffuser provide additional airflow control at low speed, high airflow conditions. The inputs to the inlet model are the flight parameters and the engine airflow command. The main outputs are the inlet forces and moments due to pressure recovery and ram and spillage drag. All resultant inlet forces and moments are resolved to body axes and modified for center of gravity offset.

The sensor component includes a rate sensor unit, an accelerometer sensor unit, an air data unit, and an inertial sensor set. The rate gyro sensors provide roll, pitch, and yaw rate measurements in the body axes. They are assumed to be perfectly accurate and have infinitely fast dynamics since the actual sensor dynamics are well beyond the control system bandwidth. The accelerometers provide acceleration measurements at the sensor locations, and they are also assumed to be perfect with infinitely fast dynamics. The air data unit measures dynamic and ambient pressure and local angles of attack and sideslip using pitot-static probes and aerodynamic vanes respectively. The inertial sensor unit provides angular orientation and inertial position and velocity measurements. Air data are augmented with inertial data to generate actual angle of attack and sideslip feedback signals. Kalman filtering is used to account for local wind effects on the air data. Again, the air data and inertial sensor units are assumed to provide perfect measurements with infinitely fast dynamics.

Dynamics of the primary structural modes are modeled to generate incremental accelerations and rotational rates at the sensor locations. The four modeled structural modes are the first wing symmetric and asymmetric modes and the first vertical and lateral fuselage modes. The inputs to the structural model are normal, lateral, and roll accelerations. Each modeled structural mode is represented by NASTRAN generated second order dynamic models.

5.1.2 Linear Model

The linear control law design process begins by generating a set of linear models at different trimmed flight conditions. As discussed in section 2.1, the nonlinear model is trimmed at a fixed flight condition and linearized by perturbing the model. The resulting linear equations representing the longitudinal and lateral/directional dynamics are given by eqs. (2.11) and (2.14), respectively. Linear models generated at different flight conditions allow the use of well established linear design methods and represent the aircraft dynamics at a range of Mach numbers and altitudes, and therefore dynamic pressures. The points chosen for linear model generation must represent a wide range of dynamic pressure, since dynamic pressure is used in the scheduling of the inner loop feedback gains.

This chapter addresses robust stabilization for the full conventional flight envelope of the modified F-18 aircraft described previously. The conventional flight envelope is defined as the subsonic Mach regime between sea level and 40,000 ft. Dynamic pressure ranges from 50 psf to 1000 psf. These limitations correspond to angles of attack up to approximately 25 deg. Fig. 5.2 represents the flight envelope considered for flight control design.

The flight conditions chosen for linear longitudinal axis design are chosen such that all linear models represent the full range of dynamic pressures for the full conventional envelope. Likewise, the lateral/directional linear models are generated to cover the same dynamic pressure range but at different Mach number and altitude combinations than those of the longitudinal models. Different design conditions are chosen for the longitudinal design and the lateral/directional design to demonstrate the lack of meaning of "design condition" for full envelope control law design. One common set of design conditions could just as easily have been chosen for both the longitudinal and lateral/directional controller designs.

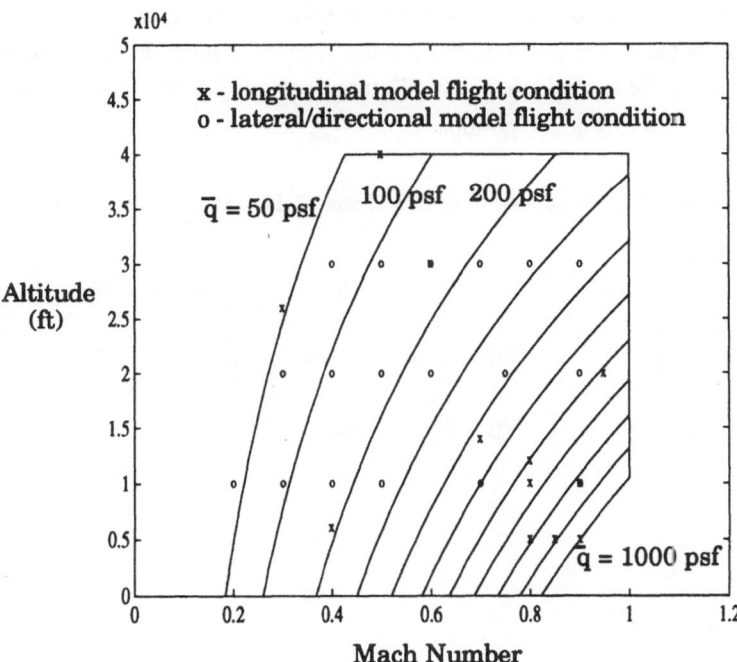

Fig. 5.2 Flight Envelope

Appendix 5 lists the flight conditions depicted in Fig. 5.2 and information about them.

5.2 Control Selector Design

As presented in section 3.1, the control selector is a mapping, given by eq. (3.2), from the generalized controls to the actual control effectors. The generalized and actual controls for the thrust vectoring vehicle are given by

$$
\delta^* = \begin{bmatrix} \dot{p}_c \\ \dot{q}_c \\ \dot{r}_c \end{bmatrix}, \qquad
\delta = \begin{bmatrix} \delta_E \\ \delta_{DT} \\ \delta_A \\ \delta_R \\ \delta_{PTV} \\ \delta_{RTV} \\ \delta_{YTV} \end{bmatrix}
\tag{5.1}
$$

where $\dot{p}$ is the roll acceleration, $\dot{q}$ is the pitch acceleration, $\dot{r}$ is the yaw acceleration, δ_E is the symmetric elevator position, δ_{DT} is the asymmetric elevator position, δ_A is the aileron position, δ_R is the rudder position, δ_{PTV} is the symmetric pitch thrust vectoring nozzle position, δ_{RTV} is the asymmetric pitch (roll) thrust vectoring nozzle position, and δ_{YTV} is the yaw thrust vectoring nozzle position.

Using eq. (5.1), the linear dynamics of eqs. (2.11) and (2.14) are written as:

$$
\begin{bmatrix} \dot{\alpha} \\ \dot{q} \\ \dot{\beta} \\ \dot{p} \\ \dot{r} \end{bmatrix} =
\begin{bmatrix}
Z_\alpha & Z_q & 0 & 0 & 0 \\
M_\alpha & M_q & 0 & 0 & 0 \\
0 & 0 & Y\beta & \sin\alpha & -\cos\alpha \\
0 & 0 & L\beta & L_p & L_r \\
0 & 0 & N\beta & N_p & N_r
\end{bmatrix}
\begin{bmatrix} \alpha \\ q \\ \beta \\ p \\ r \end{bmatrix} +
$$

$$\begin{bmatrix} Z_{\delta E} & 0 & 0 & 0 & Z_{\delta PTV} & 0 & 0 \\ M_{\delta E} & 0 & 0 & 0 & M_{\delta PTV} & 0 & 0 \\ 0 & Y_{\delta DT} & Y_{\delta A} & Y_{\delta R} & 0 & Y_{\delta RTV} & Y_{\delta YTV} \\ 0 & L_{\delta DT} & L_{\delta A} & L_{\delta R} & 0 & L_{\delta RTV} & L_{\delta YTV} \\ 0 & N_{\delta DT} & N_{\delta A} & N_{\delta R} & 0 & N_{\delta RTV} & N_{\delta YTV} \end{bmatrix} \begin{bmatrix} \delta_E \\ \delta_{DT} \\ \delta_A \\ \delta_R \\ \delta_{PTV} \\ \delta_{RTV} \\ \delta_{YTV} \end{bmatrix} \qquad (5.2)$$

The dynamics of the body-axis rates are a subset of eq. (5.2) and are given by:

$$\begin{bmatrix} \dot{p} \\ \dot{q} \\ \dot{r} \end{bmatrix} = \begin{bmatrix} 0 & 0 & L_\beta & L_p & L_r \\ M_\alpha & M_q & 0 & 0 & 0 \\ 0 & 0 & N_\beta & N_p & N_r \end{bmatrix} \begin{bmatrix} \alpha \\ q \\ \beta \\ p \\ r \end{bmatrix} +$$

$$\begin{bmatrix} 0 & L_{\delta DT} & L_{\delta A} & L_{\delta R} & 0 & L_{\delta RTV} & L_{\delta YTV} \\ M_{\delta E} & 0 & 0 & 0 & M_{\delta PTV} & 0 & 0 \\ 0 & N_{\delta DT} & N_{\delta A} & N_{\delta R} & 0 & N_{\delta RTV} & N_{\delta YTV} \end{bmatrix} \begin{bmatrix} \delta_E \\ \delta_{DT} \\ \delta_A \\ \delta_R \\ \delta_{PTV} \\ \delta_{RTV} \\ \delta_{YTV} \end{bmatrix}. \qquad (5.3)$$

Consider the following partitioning of the control effector vector:

$$\delta = \begin{bmatrix} \delta_{aero} \\ \delta_{tvec} \end{bmatrix} \quad , \text{ where } \quad \delta_{aero} = \begin{bmatrix} \delta_E \\ \delta_{DT} \\ \delta_A \\ \delta_R \end{bmatrix} \quad \delta_{tvec} = \begin{bmatrix} \delta_{PTV} \\ \delta_{RTV} \\ \delta_{YTV} \end{bmatrix}. \qquad (5.4)$$

Now eq. (5.3) becomes

$$\begin{bmatrix} \dot{p} \\ \dot{q} \\ \dot{r} \end{bmatrix} = \begin{bmatrix} 0 & 0 & L_\beta & L_p & L_r \\ M_\alpha & M_q & 0 & 0 & 0 \\ 0 & 0 & N_\beta & N_p & N_r \end{bmatrix} \begin{bmatrix} \alpha \\ q \\ \beta \\ p \\ r \end{bmatrix} + [B_{aero} \quad B_{tvec}] \begin{bmatrix} \delta_{aero} \\ \delta_{tvec} \end{bmatrix}. \qquad (5.5)$$

where

$$B_{aero} = \begin{bmatrix} 0 & L_{\delta DT} & L_{\delta A} & L_{\delta R} \\ M_{\delta E} & 0 & 0 & 0 \\ 0 & N_{\delta DT} & N_{\delta A} & N_{\delta R} \end{bmatrix},$$

$$B_{tvec} = \begin{bmatrix} 0 & L_{\delta RTV} & L_{\delta YTV} \\ M_{\delta PTV} & 0 & 0 \\ 0 & N_{\delta RTV} & N_{\delta YTV} \end{bmatrix}, \tag{5.6}$$

and δ_{aero} and δ_{tvec} are given by eq. (5.4). To preserve the dynamics of the body-axis rates, eq. (3.1) is written as:

$$[B_{aero} \ \ B_{tvec}] \begin{bmatrix} \delta_{aero} \\ \delta_{tvec} \end{bmatrix} = B^* \delta^*, \text{ where } \qquad B^* = \begin{bmatrix} 1 & 0 & 0 \\ 0 & 1 & 0 \\ 0 & 0 & 1 \end{bmatrix}, \tag{5.7}$$

δ^* is given by eq. (5.1), and the generalized control effectiveness matrix, B^*, is identity since the generalized controls correspond to the rotational degrees-of-freedom accelerations.

A daisy-chain method is used to generate thrust vector commands. Thrust vectoring is used only when the aerodynamic surfaces are not able to generate the necessary forces and moments required for commanded maneuvers. Therefore, the computation of aerodynamic control commands is independent of thrust vectoring control commands. The control selector is then defined using eqs. (3.2) and (3.3)

$$\delta_{aero} = T_{aero}\,\delta^*, \qquad\qquad\qquad \delta_{tvec} = T_{tvec}\,\delta^* \tag{5.8}$$

and

$$T_{aero} = N_{aero}(B_{aero}N_{aero})^\#, \quad T_{tvec} = N_{tvec}(B_{tvec}N_{tvec})^\#, \tag{5.9}$$

where N_{aero} and N_{tvec} are used to prioritize the use of redundant control effectors. Since the ailerons contribute more to the roll acceleration and the first priority of the horizontal tail should be pitch control, the differential horizontal tail command is reduced by weighting the command to be a quarter of the other aerodynamic commands. There is no redundancy for the thrust vectoring control effectors, and thus, the prioritization matrices become

$$N_{aero} = \begin{bmatrix} 1 & 0 & 0 & 0 \\ 0 & .25 & 0 & 0 \\ 0 & 0 & 1 & 0 \\ 0 & 0 & 0 & 1 \end{bmatrix} \qquad N_{tvec} = \begin{bmatrix} 1 & 0 & 0 \\ 0 & 1 & 0 \\ 0 & 0 & 1 \end{bmatrix}. \tag{5.10}$$

Computation of the control selector equation (5.9) depends on flight condition. Therefore, the elements of B_{aero} and B_{tvec} are found using linear interpolation of stored table values.

Nonlinear elements, such as position and rate limits, are required to implement the daisy-chain thus making the control selector nonlinear. Fig. 5.3 shows the structure of the nonlinear control selector.

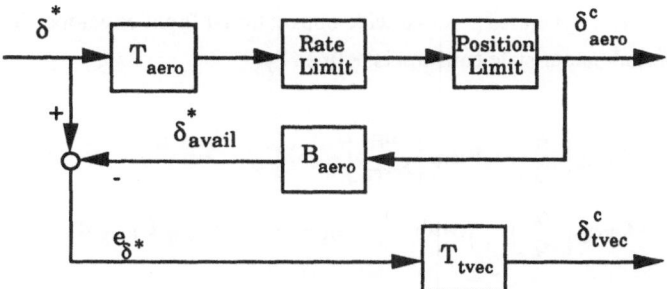

Fig. 5.3 Nonlinear Control Selector

A limited aerodynamic surface command (δ^c_{aero}) is generated from a rotational acceleration command (δ^*) via the aerodynamic control selector (T_{aero}) and aerodynamic surface rate and position limits. An achievable rotational acceleration vector (δ^*_{avail}) is computed from the limited aerodynamic surface command using the aerodynamic control distribution (B_{aero}). The difference of the commanded and achievable rotational acceleration vectors (e_{δ^*}) is transformed to a thrust vector command (δ^c_{tvec}) using the thrust vector control selector (T_{tvec}).

The bottom line of transformation to generalized controls is seen by combining the state dynamics and the control selector. The resulting longitudinal and lateral/directional linear state dynamical equations are:

$$\begin{bmatrix} \dot\alpha \\ \dot q \end{bmatrix} = \begin{bmatrix} Z_\alpha & Z_q \\ M_\alpha & M_q \end{bmatrix} \begin{bmatrix} \alpha \\ q \end{bmatrix} + B^*_{long}\, \dot q_c \qquad \text{where } B^*_{long} = \begin{bmatrix} 0 \\ 1 \end{bmatrix}, \qquad (5.11)$$

and

$$\begin{bmatrix} \dot\beta \\ \dot p \\ \dot r \end{bmatrix} = \begin{bmatrix} Y\beta & \sin\alpha & -\cos\alpha \\ L\beta & L_p & L_r \\ N\beta & N_p & N_r \end{bmatrix} \begin{bmatrix} \beta \\ p \\ r \end{bmatrix} + B^*_{lat} \begin{bmatrix} \dot p_c \\ \dot r_c \end{bmatrix} \text{ where } B^*_{lat} = \begin{bmatrix} 0 & 0 \\ 1 & 0 \\ 0 & 1 \end{bmatrix}, \quad (5.12)$$

based on the above control selector definition. Therefore, eqs. (5.11) and (5.12) represent
the form of the models used for linear control design.

5.3 Longitudinal Axis Controller

The longitudinal axis manual flight control system generates elevator and pitch thrust vector
commands from angle of attack commands. These control effector commands stabilize the
short period aircraft dynamics across a wide conventional flight envelope. The open-loop
linear short period output-state equation is defined by:

$$\begin{bmatrix} \dot{\alpha} \\ \dot{q} \end{bmatrix} = A_{long} \begin{bmatrix} \alpha \\ q \end{bmatrix} + B_{long} \begin{bmatrix} \delta_E \\ \delta_{PTV} \end{bmatrix}$$

$$\begin{bmatrix} \alpha \\ q \end{bmatrix} = C_{long} \begin{bmatrix} \alpha \\ q \end{bmatrix} \quad \text{with} \quad G_{long} \equiv (A_{long}, B_{long}, C_{long}, 0), \qquad (5.13)$$

where A_{long} and B_{long} are defined in (2.11), and C_{long} is obviously identity. Taking into
account the function of the control selector, eq. (5.13) becomes

$$\begin{bmatrix} \dot{\alpha} \\ \dot{q} \end{bmatrix} = A_{long} \begin{bmatrix} \alpha \\ q \end{bmatrix} + B^*_{long} \dot{q}_c$$

$$\begin{bmatrix} \alpha \\ q \end{bmatrix} = C_{long} \begin{bmatrix} \alpha \\ q \end{bmatrix} \quad \text{with} \quad \tilde{G}_{long} \equiv (A_{long}, B^*_{long}, C_{long}, 0), \qquad (5.14)$$

where B^*_{long} is defined in eq. (5.11).

A minimal-order H_∞ design algorithm is used for the design of an inner loop
equalization controller. Structured singular value synthesis is used to design outer loop
robust performance controllers. Different control laws are found for high and low dynamic
pressure conditions, and controller commands are blended for a small region of dynamic
pressure. The inner and outer loop control designs are presented followed by robustness
and flying qualities analyses for the entire longitudinal axis flight controller.

5.3.1 Inner Loop Design

The structure of the aircraft with a closed inner loop controller is shown in Fig. 5.4.

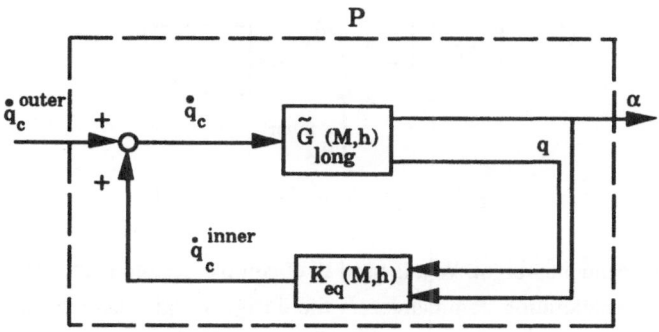

Fig. 5.4 Inner Loop Equalization

The pitch acceleration command, $\dot{q}_c$, is made up of an inner loop pitch acceleration command, $\dot{q}_c^{inner}$, and an outer loop pitch acceleration command, $\dot{q}_c^{outer}$. P is a linear model of the inner equalization controller, K_{eq}, and a combination of the open-loop aircraft dynamics and the control selector, $\tilde{G}_{long}$. Dependency of an element on flight condition is represented by (M,h).

The purpose of the inner loop equalization controller is, as the name suggests, to equalize the transfer function between angle of attack response and outer loop pitch acceleration command across the flight envelope. In other words, the goal of the inner loop is to reduce the variation of the aircraft dynamics between operating conditions, thereby reducing the modeling uncertainty between flight condition dependent aircraft models. Successfully reducing the modeling uncertainty provides a greater possibility that one robust performance controller will provide stability and performance across all operating conditions. Therefore an effective inner loop controller must be designed to reduce the relative error between P and P_0 as defined by eq. (3.4).

Since P is a function of flight condition, the equalization module must be scheduled against slowly-varying flight condition-dependent parameters [5.2], such as dynamic pressure, altitude, or Mach number. Gain scheduling of the inner loop controller is necessary because the aircraft dynamics vary greatly across a wide envelope, making it impossible to design a single robust controller. In this design, the inner loop equalization controller gains are scheduled across the flight envelope as functions of dynamic pressure.

The minimal-order H_∞ design algorithm described in section 2.8 is used to design a longitudinal axis inner loop equalization controller for the design model shown in Fig. 5.5.

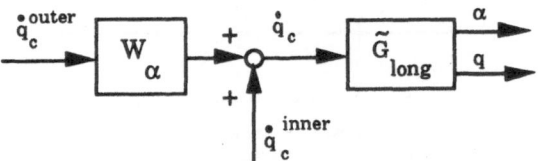

Fig. 5.5 Inner Loop Design Model

A frequency dependent weight, W_α, is used to obtain the desired response of the angle of attack to pitch acceleration commands. If the design weight has the realization $W_\alpha \equiv (A_W, B_W, C_W, D_W)$, then the parameters of the design model from Fig. 2.10 become

$$A = \begin{bmatrix} A_{long} & B_{long}^* C_W \\ 0 & A_W \end{bmatrix}, \qquad B = \begin{bmatrix} B_{long}^* \\ 0 \end{bmatrix}, \qquad C = [\ C_{long}\ \ 0\ ],$$

$$G_1 = \begin{bmatrix} B_{long}^* D_W \\ B_W \end{bmatrix}, \qquad H_1 = [\ C^*\ \ 0\ ],$$

$$u = \dot{q}_c^{inner}, \qquad x = y = \begin{bmatrix} \alpha \\ q \end{bmatrix}, \qquad w_1 = \dot{q}_c^{outer}, \qquad z_1 = \alpha, \tag{5.15}$$

and H_2 is the control signal weighting used as a design parameter. Since the control system will eventually track angle of attack commands, the performance output distribution matrix is

$$C^* = [\ 1\ \ 0\ ]. \tag{5.16}$$

For the flight conditions of Fig. 5.2, the magnitudes of the open loop pitch acceleration to angle of attack transfer functions are shown in Fig. 5.6. There is a wide variation in aircraft dynamics across the flight envelope. The high dynamic pressure flight condition models have lower magnitudes at low frequency than the low dynamic pressure flight condition models. Since the purpose of the inner loop controller is to equalize these dynamics, the design weight, W_α, is chosen as the inverse of some transfer function that lies among the span of open loop dynamics. It is suggested that the inverse design weight dynamics be chosen close to the dynamics of the high dynamic pressure flight conditions so that positive feedback is not used to diminish the fast dynamics of the high dynamic pressure conditions.

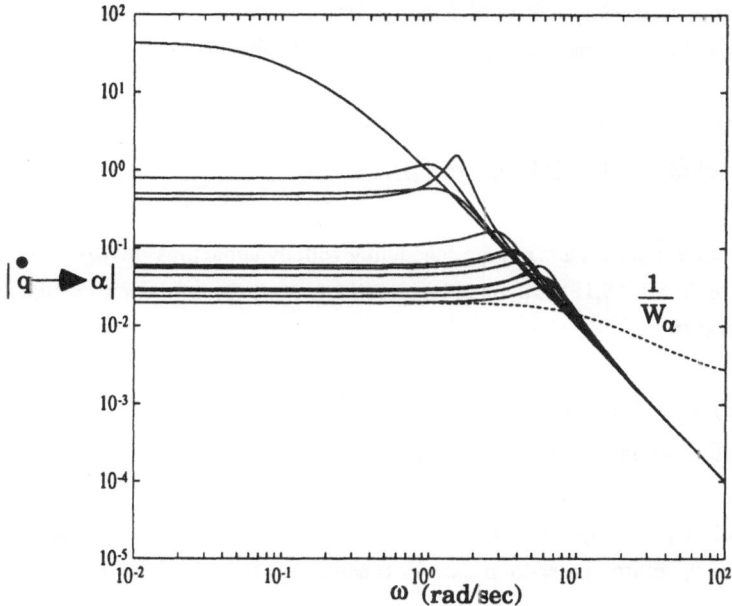

Fig. 5.6 Pitch Acceleration to Angle of Attack Open Loop Dynamics

With the performance weight augmented to the open-loop aircraft dynamics, the number of design states becomes three, and the number of inputs and outputs remains at one and two respectively. Therefore, the order of the equalization controller is one (# of states - # of measurements). The performance and control signal weights used for the inner equalization design are

$$A_W = -100 \qquad B_W = 1 \qquad C_W = -45918 \quad D_W = 510.20$$
$$H_{\hat{z}} = .035 . \tag{5.17}$$

Since the purpose of the inner loop controller is to equalize the dynamics across the flight envelope, the performance weight W_α, given by eq. (5.17), is chosen as the inverse of a first order fit to the open-loop dynamics at the highest dynamic pressure design condition in Fig. 5.2. The choice of the highest dynamic pressure dynamics prevents positive feedback from diminishing the fast dynamics at high dynamic pressure conditions which will inevitably occur for other choices of W_α.

Controllers are designed for each condition in Fig. 5.2, and some controller parameters are scheduled with polynomial fits in dynamic pressure. The resulting controller parameters, with design parameter $a = 1$ and design weights given by eq. (5.17), are

$$F = -40 \qquad K_f = [1 \quad 1] \qquad G = 0.0247$$
$$N = N(\bar{q}) \qquad M = [M_1(\bar{q}) \quad M_2(\bar{q})]. \tag{5.18}$$

Only three control parameters require scheduling with dynamic pressure $(\bar{q})$. The elements of N and M in eq. (5.18) are fit using least-squares polynomial curve-fits, linear in dynamic pressure:

$$N(\bar{q}) = -.312 \, \bar{q} + 461$$
$$M_1(\bar{q}) = -.058 \, \bar{q} + 50.5 \qquad M_2(\bar{q}) = -.006 \, \bar{q} + 8.11 \tag{5.19}$$

The controller given by eqs. (5.18) and (5.19) provides closely matched frequency responses of the closed inner-loop transfer function (P) from $\dot{q}_c^{outer}$ to α across the flight envelope as shown in Fig. 5.7.

It is assumed that equalizing the pitch acceleration to angle of attack dynamics will also equalize the pitch acceleration to pitch rate dynamics since they are not independent. The open loop pitch acceleration to pitch rate transfer functions are shown in Fig. 5.8. Closing the inner loop with the controller given by eqs. (5.18) and (5.19) approximately equalizes the pitch acceleration to pitch rate dynamics as shown in Fig. 5.9.

To analyze the variation of aircraft dynamics at different flight conditions, the relative error defined by eq. (3.4) is computed for a given linear model (P) and a nominal linear model (P_0). In order to evaluate the relative error between the equalized models of Fig. 5.7, the closed inner-loop model (P) at Mach=0.95 and altitude=20 kft is arbitrarily chosen as the nominal model (P_0). Fig. 5.10 shows the maximum singular values of the model errors relative to P_0 across the flight envelope at design conditions given in Fig. 5.2. Since the closed inner-loop errors relative to the Mach=0.95 and altitude=20kft flight condition are less than unity for all frequencies and across the flight envelope, P_0 is used as the open outer-loop design plant for the robust performance controller.

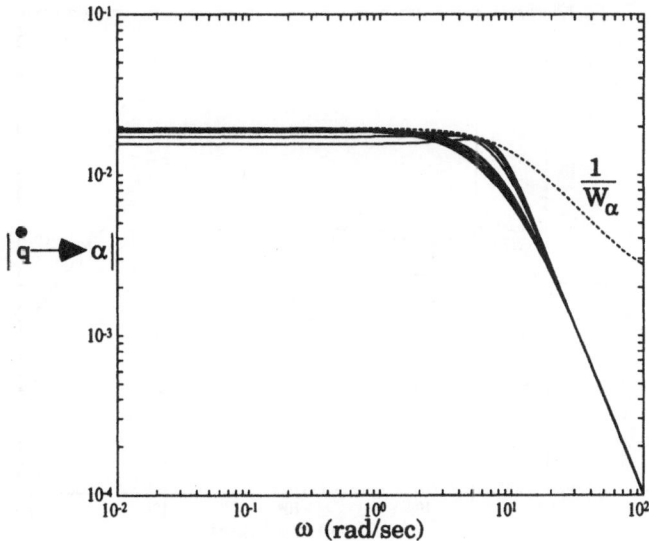

Fig. 5.7 Pitch Acceleration to Angle of Attack Closed Inner-Loop Dynamics

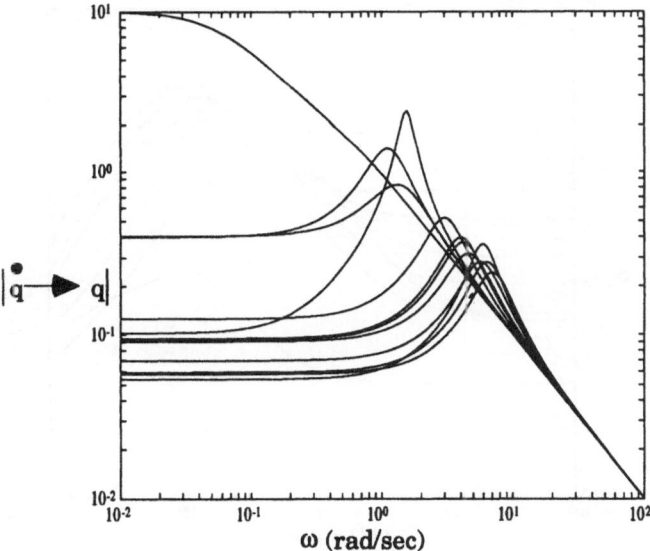

Fig. 5.8 Pitch Acceleration to Pitch Rate Open Loop Dynamics

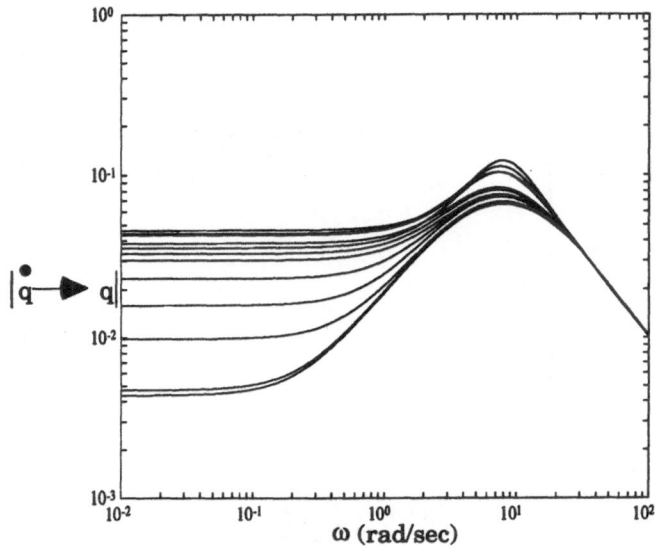

Fig. 5.9 Pitch Acceleration to Pitch Rate Closed Inner-Loop Dynamics

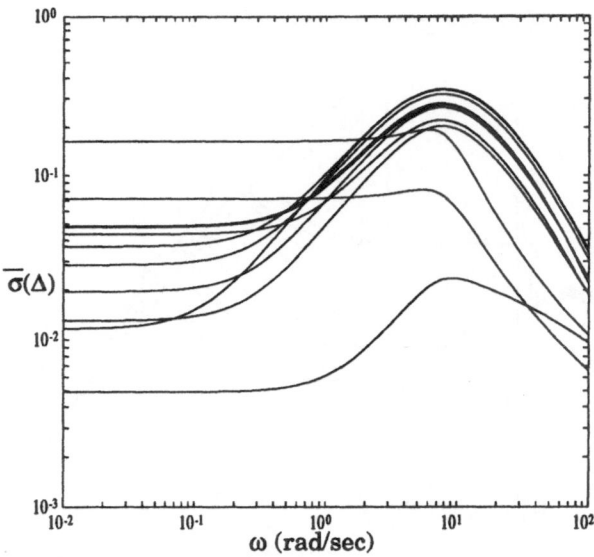

Fig. 5.10 Closed Inner-Loop Relative Errors

5.3.2 Outer Loop Design

The structure of the system with the outer loop controller is shown in Fig. 5.11.

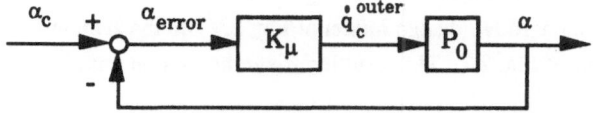

Fig. 5.11 Outer Loop Control

P_0 is the closed inner loop nominal model (or the open outer loop model), and the robust outer loop performance controller is K_μ. The purpose of the outer loop performance controller is to generate a pitch acceleration command ($\dot{q}_c^{outer}$), from an angle of attack command (α_c), that produces a desired robust aircraft angle of attack response (α).

This problem is formulated as an implicit model-following problem [5.3], where the ideal model to be followed is chosen to be a second-order transfer function

$$\frac{\alpha_{ideal}}{\alpha_c} = \frac{\omega_{ideal}^2}{s^2 + 2\,\zeta_{ideal}\,\omega_{ideal}\,s + \omega_{ideal}^2}. \tag{5.20}$$

with ζ_{ideal} and ω_{ideal} chosen based on flying qualities requirements from section 2.2. Using this ideal model-following approach, the outer loop performance controller design model is developed in Fig. 5.12.

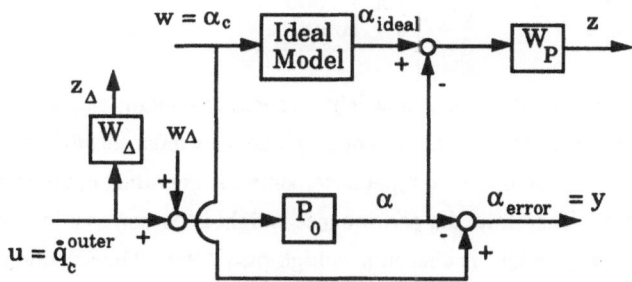

Fig. 5.12 Outer Loop Design Model

P_0 is the design plant described previously, and W_P and W_Δ are performance and input uncertainty design weights, respectively.

Structured singular value (μ) synthesis is used to design outer loop performance controllers for the system in Fig. 5.12. As described in section 2.9,
μ-synthesis is a combination of general H_∞ feedback design and structured singular value analysis. Therefore, the design model of Fig. 5.12 is organized into the
structure of the standard H_∞ design model of Fig. 2.9 and the standard structured singular value M-Δ format of Fig. 2.7. The resulting μ-synthesis model is shown in Fig. 5.13.

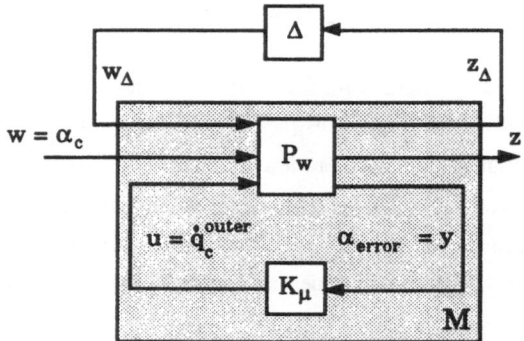

Fig. 5.13 Structured Singular Value Synthesis Model

P_W is the weighted outer loop design plant, or the weighted P_0.

The robust performance controller (K_μ) for P_W is designed using μ-synthesis with the performance and input uncertainty weights set as follows:

$$W_P = \frac{.25s + 50}{s + 5} \qquad W_\Delta = \frac{10s + 1000}{s + 10000}. \tag{5.21}$$

The process of choosing these design weights is somewhat empirical. The tracking of the ideal angle of attack response is important at frequencies most sensitive to the pilot (low frequencies), whereas robustness to input uncertainties is important at different frequencies (high frequencies). Therefore, the performance weight is chosen as a low-pass filter, and the input uncertainty weight is chosen as a high-pass filter. The selection of the poles, zeros, and gains of the filters is iterative. Initial filter values are chosen, a controller is designed, and time responses are examined. The filter values are tuned until satisfactory tracking occurs with minimal control effort.

Unlike responses in the lateral/directional axes, the desired pitch response is not uniform across the flight envelope. The pilot would like to feel a faster response at higher speeds. To account for the different performance requirements, two robust performance controllers are designed using two different ideal models based on eq. (5.20): 1) a low dynamic pressure controller using a slow ideal model (ζ_{ideal}=0.8, ω_{ideal}=3 rad/sec), and 2) a high dynamic pressure controller using a fast ideal model (ζ_{ideal}=0.8, ω_{ideal}=5 rad/sec). The resulting controllers are 13th order and of the following form:

$$K_\mu(s) = \frac{K(s + z_1)(s + z_2)\cdots(s + z_{12})(s + z_{13})}{(s + p_1)(s + p_2)\cdots(s + p_{12})(s + p_{13})} ,$$

(5.22)

where the numerator and denominator coefficients are given in Appendix 5. Implementation of the two outer-loop performance controllers is discussed later in this section.

As seen by eq. (5.22), μ-synthesis typically generates controllers of an order much higher than the original plant because of the design weights and the frequency dependent scalings discussed in section 2.9. The additional state variables in the controllers are a result of the design method. Therefore, good performance is possible with a reduced-order controller. The structured singular values of the closed loop system of Fig. 5.13 are computed with the 13th-order controller implemented. Fig. 5.14 and Fig. 5.15 show that the maximum singular values for the low and high dynamic pressure controllers are 0.95 and 0.85, respectively. The balanced truncation method of section 2.10 is used to reduce the controller order without degrading the performance and robustness of the controller. A maximum number of transformed controller states, z, of eq. (2.81), are truncated such that the maximum structured singular value does not degrade above unity for the closed-loop system of Fig. 5.13. The controller is reduced from 13th order to 4th order without significant performance or robustness deterioration. Therefore, using the notation of section 2.10, the dimension, k, of the truncated transformed controller state vector, z_2, is 9, and the dimension of the reduced-order state vector, z_1, is 4. Fig. 5.14 shows the structured singular values of the closed-loop system with full- and reduced-order low dynamic pressure controllers.

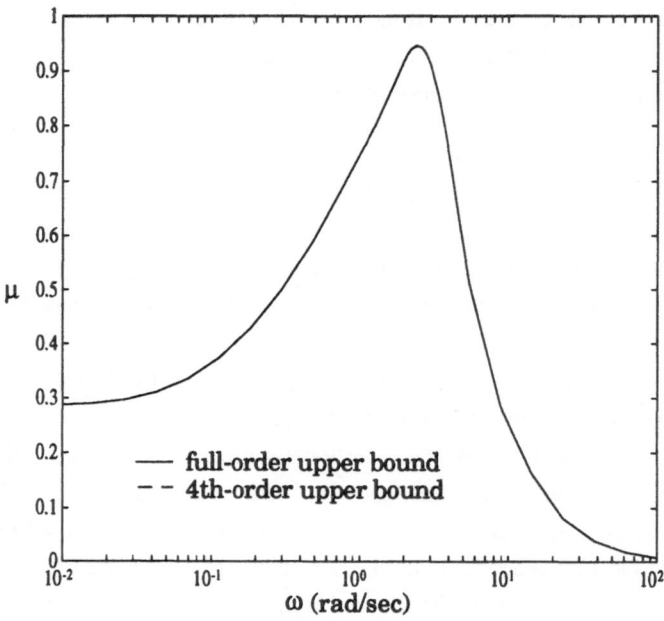

Fig. 5.14 Closed-Loop System with Low $\bar{q}$ Controller

It is seen that reduction of the full 13th-order controller to a 4th-order controller does not destroy the original robustness and performance. Fig. 5.15 shows similar results for the high dynamic pressure controller. The final 4th-order low and high dynamic pressure controllers are

$$K_\mu^{lo}(s) = \frac{1.30 \times 10^2 (s + 6.02 \times 10^2)(s + 3.84 \times 10^0 \pm j6.25 \times 10^0)}{(s + 5.91 \times 10^1 \pm j5.20 \times 10^1)(s + 3.63 \times 10^{-2})(s + 6.64 \times 10^0)}$$

$$(5.23)$$

$$K_\mu^{hi}(s) = \frac{1.20 \times 10^2 (s + 2.42 \times 10^3)(s + 5.40 \times 10^0 \pm j7.26 \times 10^0)}{(s + 7.18 \times 10^1 \pm j4.05 \times 10^1)(s + 7.23 \times 10^{-2})(s + 2.05 \times 10^0)} \cdot$$

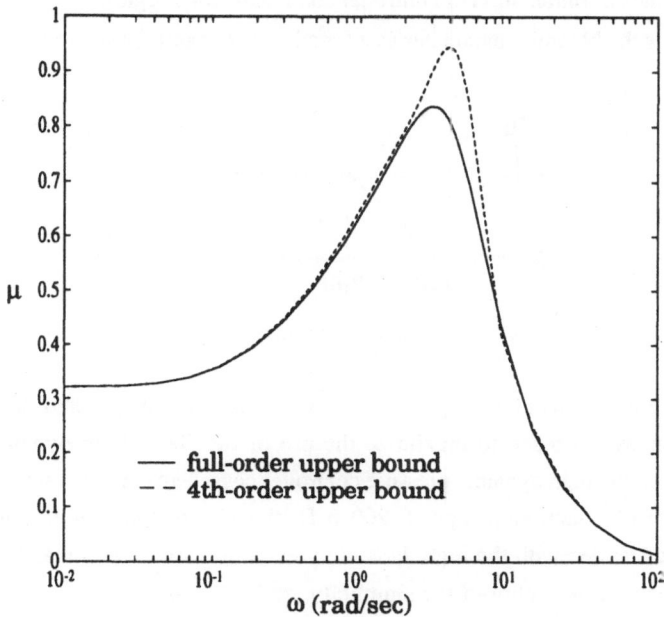

Fig. 5.15 Closed-Loop System with High $\bar{q}$ Controller

It is known that at high dynamic pressure conditions, K_μ^{hi} should be used, and at low dynamic pressure conditions, K_μ^{lo} should be used. However, the definition of high and low dynamic pressure is unclear. The multiple dynamic compensators of eq. (5.23) are implemented through a blending of controller commands over a range of dynamic pressure. Fig. 5.16 schematically shows the combination of controller commands generated from the high and low dynamic pressure controllers of eq. (5.23).

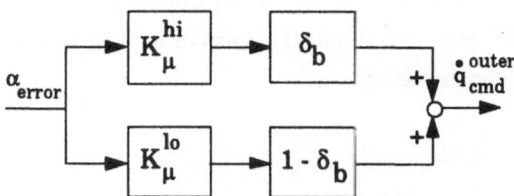

Fig. 5.16 Outer Loop Controller Blending

Implementation of the controllers is achieved through a simple linear blending parameter (δ_b) that generates a combination of controller commands for a region of dynamic pressure. Fig. 5.17 shows the blending parameter as a function of dynamic pressure.

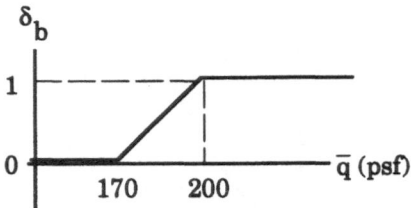

Fig. 5.17 Blending Parameter Function

The dynamic pressure blending region between 170 psf and 200 psf and the piece-wise linear function were chosen to maximize the use of the "fast" high dynamic pressure controller. Once the high dynamic pressure controller caused an elevator rate saturation for a 5 deg. angle of attack step input (~200 psf), the low dynamic pressure controller command was blended with the high dynamic pressure controller command. The piece-wise linear function was chosen for simplicity, and a blending region of about 30 psf seemed to minimize the transient blending effects in nonlinear simulations.

5.3.3 Robustness Analysis

Structured singular value analysis techniques are used to analyze stability robustness of the designed control system to uncertainties corresponding to unmodeled actuator and sensor dynamics, parameters in the plant model, and blending of high and low dynamic pressure outer performance controllers.

The reduced order actuator models, given in Appendix 5, are used in the nonlinear model and the linear analysis model, and the difference between the high-order and low-order models represents the unmodeled actuator dynamics [5.1]. The actual error dynamics between these models are fit to real rational transfer functions that become the weighting functions for the actuator uncertainty

$$W_E = \frac{.63s^2 + 3.03s + .078}{s^2 + 68.4s + 1900}.$$
(5.24)

W_E is the elevator uncertainty weight. The absence of a pitch thrust vector uncertainty weight is due to the fact that thrust vectoring is used only during aerodynamic effector saturation. Since saturation is a nonlinear phenomenon, linear robustness analysis is inappropriate.

The angle of attack and pitch rate sensor dynamics are captured entirely as unstructured uncertainty. The sensor dynamics are estimated from flight test data of high performance aircraft [5.1] and fit to real rational transfer functions representing measurement uncertainty weighting functions

$$W_{\alpha} = \frac{21.9s^2 + 1120s + 91100}{s^2 + 574s + 1140000}$$

$$W_q = \frac{.745s^3 + 152s^2 + 95.9s + 1.38}{s^3 + 626s^2 + 173000s + 235000}.$$

$$(5.25)$$

W_{α} is the angle of attack sensor uncertainty weight, and W_q is the pitch rate sensor uncertainty weight.

An extensive aerodynamic uncertainty database, developed from wind tunnel and flight test data [5.1], is used to generate structured uncertainty models of stability and control derivatives. These models are then translated into state-space element uncertainty models. The state-space element uncertainties depend on flight condition. However for simplicity, the uncertainty model for each state-space element,

$$\Delta A = \begin{bmatrix} \pm.02|Z_{\alpha}| & 0 \\ \pm.04|M_{\alpha}| & \pm2|M_q| \end{bmatrix}$$

$$\Delta B = \begin{bmatrix} \pm.22|Z_{\delta E}| \\ \pm.04|M_{\delta E}| \end{bmatrix},$$

$$(5.26)$$

is held constant at the worst flight condition case thus making the parameter uncertainty models conservative in a sense. The additive uncertainty structure enters the state equation as

$$\dot{x} = (A + \Delta A)x + (B + \Delta B)u .$$

$$(5.27)$$

Recall from section 5.3.2 that for better flying qualities, high and low dynamic pressure robust performance controllers are designed and implemented by blending both controller commands with the blending parameter δ_b. The actual blending parameter, δ_b, varies between 0 and 1. However, it is normalized such that the blending parameter for analysis,

$\bar{\delta}_b$, varies between -1 and 1. Therefore, analyzing robustness to $\bar{\delta}_b$ is equivalent to analyzing robustness for all outer-loop controller command combinations.

Fig. 5.18 shows the robustness analysis model including the structure of the uncertainty as well as the normalization weighting elements. To show the uncertainty structure, note that the diagram is broken down to the scalar loop level except for the plant parameter uncertainty loops whose structure is given by eq. (5.27).

Fig. 5.19 shows the robust stability plots of the structure shown in Fig. 5.18 for all robustness analysis conditions given in Appendix 5. The robust stability plots are the mixed real/complex structured singular values. It is interesting to note the absence of a peak near the short period frequency. This suggests insensitivity of the short period dynamics to the given uncertainty structure. The peak at approximately 10 rad/sec corresponds to sensitivity to elevator actuation. If a redesign of the control system were within the scope of this task, inclusion of an elevator actuation weight in the design model might increase robustness.

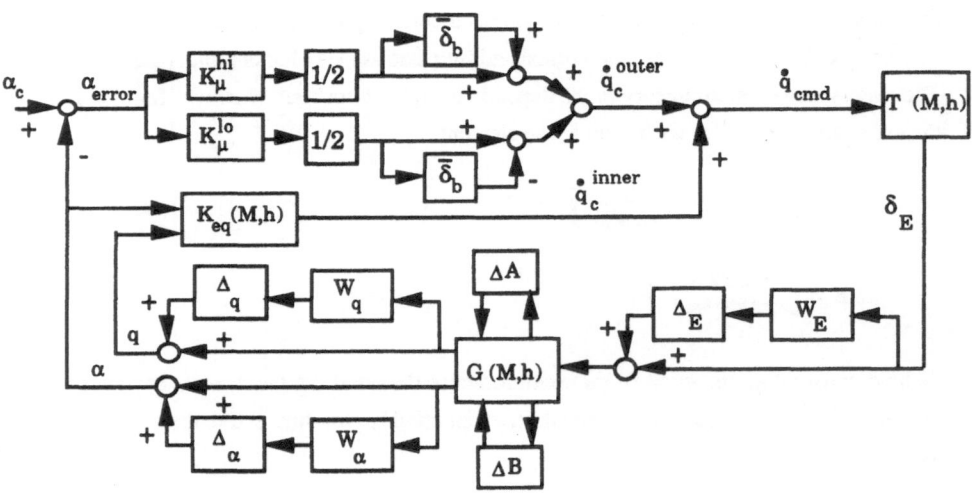

Fig. 5.18 Stability Robustness Analysis Model

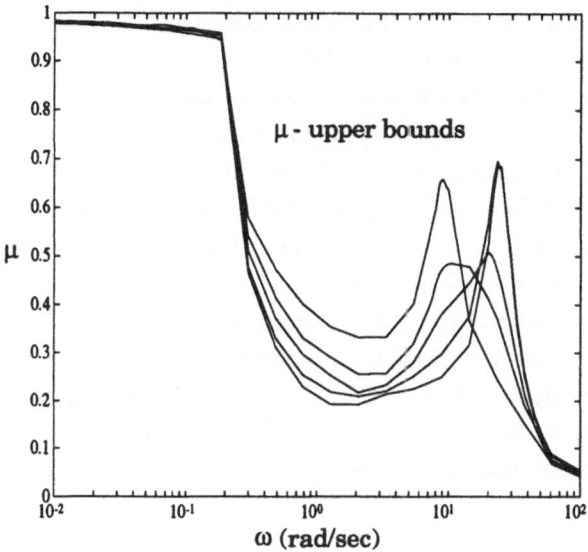

Fig. 5.19 Robust Stability

5.3.4 Flying Qualities Analysis

Several flying qualities measures of section 2.2 are used to analyze the handling characteristics of the aircraft and the designed control system. The measures for pitch response are given in terms of low-order equivalent systems (LOES) that represent pitch rate (q) and normal acceleration at the instantaneous center of rotation n_z' responses to pilot stick deflection inputs (δ_p). Only considering the short-term pitch response from eq. (2.10), the LOES become

$$\frac{q}{\delta_p} = \frac{K_\theta s\left(s + \dfrac{1}{T_{\theta 2}}\right)e^{-\tau_\theta s}}{s^2 + 2\zeta_{sp}\omega_{sp}s + \omega_{sp}^2} \qquad \frac{n_z'}{\delta_p} = \frac{K_n e^{-\tau_n s}}{s^2 + 2\zeta_{sp}\omega_{sp}s + \omega_{sp}^2}, \tag{5.28}$$

where ζ_{sp} is equivalent short-period damping, ω_{sp} is equivalent short-period frequency, $T_{\theta 2}$ corresponds to the pitch rate zero , and τ_θ and τ_n are equivalent pitch time delays. Only the parameters of the stick to pitch rate transfer function are considered for analysis since once $T_{\theta 2}$ is defined and it is assumed the time delays are equal, the transfer functions only differ by the steady-state gain. The complete scheduled flight control system, full-order actuator models, and second-order aircraft short-period models are used to generate high-order closed-loop aircraft linear models. LOES are generated from these high-order models using an equivalent system transfer function matching program.

The Control Anticipation Parameter (CAP) defined by eq. (2.18) is computed using the appropriate LOES parameters, and Fig. 5.20 shows that all flight conditions meet Category A, Level 1 flying qualities requirements for CAP and ω_{sp} except at the Mach 0.4, 6 kft flight condition which is slightly below Level 1. Figs. 5.21 and 5.22 show that Level 1 requirements are met for ω_{sp}, $T_{\theta 2}$, ζ_{sp}, and τ_{θ}.

The mismatch defined by eq. (2.15), ranges from 0.009 to 4.376 which suggests a close LOES approximation. Section 2.2 describes the mismatch as well as maximum unnoticeable added dynamic bounds. Figs. 5.23 and 5.24 show that the stick deflection to pitch rate added high order dynamics are well within the maximum allowable unnoticeable added dynamics represented by the dashed lines.

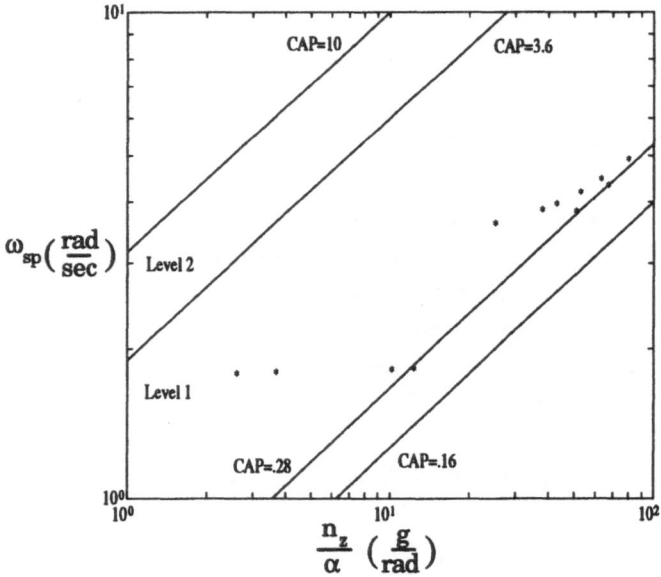

Fig. 5.20 Control Anticipation Parameter

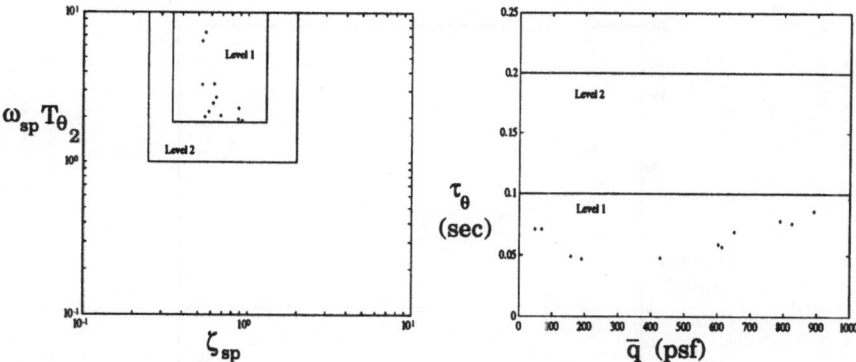

Fig. 5.21 Damping and Frequency   Fig. 5.22 Equivalent Time Delay

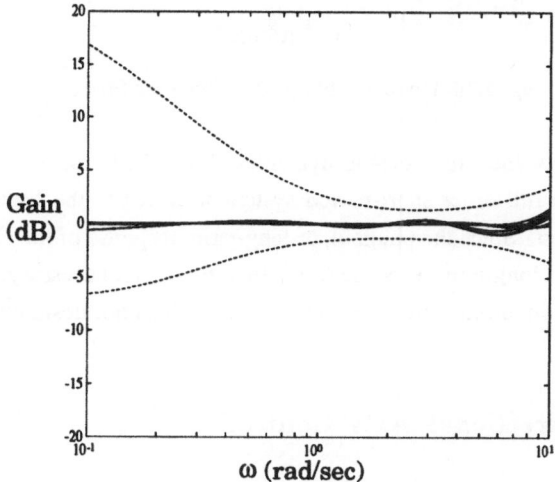

Fig. 5.23 Unnoticeable Added Dynamic Gain

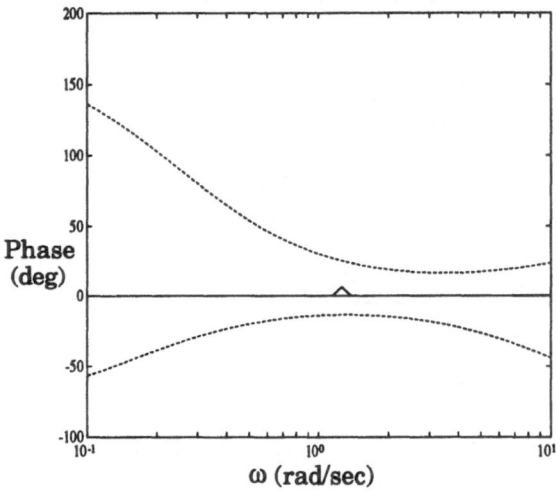

Fig. 5.24 Unnoticeable Added Dynamic Phase

The mismatch values and unnoticeable dynamics, Figs. 5.23 and 5.24, show that the LOES estimates the full-order short-period system well across the flight envelope for frequencies of interest. Since the phugoid, or long-term, response of the aircraft is stable, a detailed analysis of long-term response flying qualities is not necessary. The nonlinear time responses in section 5.5 show that control system does not destabilize the phugoid mode.

5.4   Lateral/Directional Axis Controller

The lateral/directional axes manual flight control system generates differential elevator, aileron, rudder, and roll and yaw thrust vector commands from sideslip and stability-axis roll rate commands. These control effector commands stabilize the roll and Dutch roll aircraft dynamics across a wide conventional flight envelope. A robust eigenstructure assignment design algorithm is used for the design of the inner loop equalization controller. Structured singular value synthesis is used to design outer loop robust performance controllers. The inner and outer loop control designs are presented followed by robustness and flying qualities analyses for the entire lateral/directional axes flight controller.

5.4.1 Inner Loop Design

The desired eigenstructure for the inner loop is chosen to consist of a real pole and a complex pair of poles that correspond to the desired equivalent roll and Dutch roll poles from the desired flying qualities. The desired eigenvalues are

$$\lambda_1{}^d = -1.67 \quad \text{and} \quad \lambda_{2,3}{}^d = -2.1 \pm 2.14\, j \, . \tag{5.29}$$

The desired eigenvectors corresponding to the desired eigenvalues are

$$v_1{}^d = \begin{bmatrix} 1 \\ x \\ 0 \end{bmatrix} \quad v_2{}^d = \begin{bmatrix} 0 \\ x \\ 1 \end{bmatrix} \quad v_3{}^d = \begin{bmatrix} 0 \\ 1 \\ x \end{bmatrix}, \tag{5.30}$$

where x represents the unspecified elements in the eigenvectors. Since the control effectiveness matrix is replaced by the generalized control effectiveness matrix, the plant is already in the form of eq. (2.54). The gain matrix from the three outputs to the two generalized inputs is found for each flight condition. The gain from the roll rate to the generalized yaw acceleration command is constrained to be zero, since the gain would otherwise be insignificant. This element can be constrained to be zero directly in the equations used to find the feedback gain matrix. The feedback matrix eq. (2.58) is rewritten in terms of a Kronecker product and a row stacking operator, so the elements of the feedback gain matrix are written in terms of a vector. Each row is solved independently for each of the feedback matrix elements, and any element can be constrained to be zero by removing the corresponding row of the matrix equation, and by removing that element from the vector of elements.

Each of the elements in the inner loop gain matrix is plotted with dynamic pressure, and a polynomial is found to approximate the gains at each of the points. The yaw rate to roll acceleration and yaw acceleration gains are also functions of altitude, so they are scheduled as a function of static pressure ratio as well. For a particular dynamic pressure, the gain is found at discrete altitudes and interpolated to find the specific gain at the corresponding altitude. The plots of the inner loop gains appear in Fig. 5.25. The three curves in Figs. 5.25c and 5.25e represent the gains at three different altitudes.

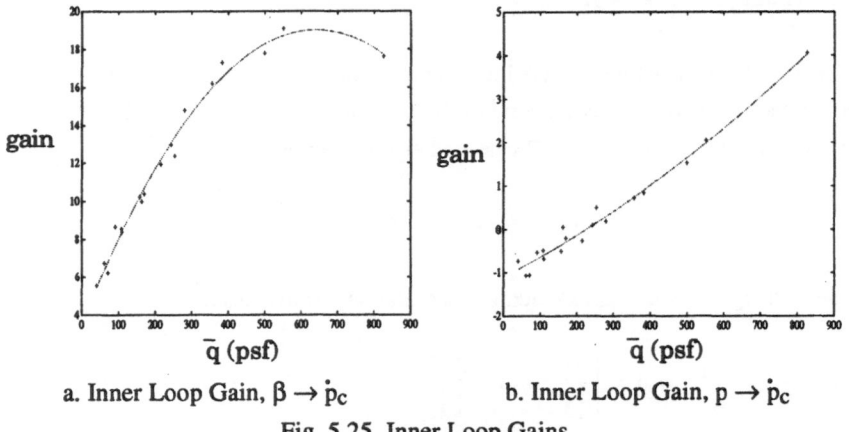

a. Inner Loop Gain, $\beta \rightarrow \dot{p}_c$ b. Inner Loop Gain, $p \rightarrow \dot{p}_c$

Fig. 5.25 Inner Loop Gains

The polynomial gain functions of Fig. 5.25 are of the following form:

$$\text{gain} = c_2 \, \bar{q}^2 + c_1 \, \bar{q} + c_0 \,, \tag{5.31}$$

where the coefficients c_i, i=0,1,2, are given in Appendix 5.

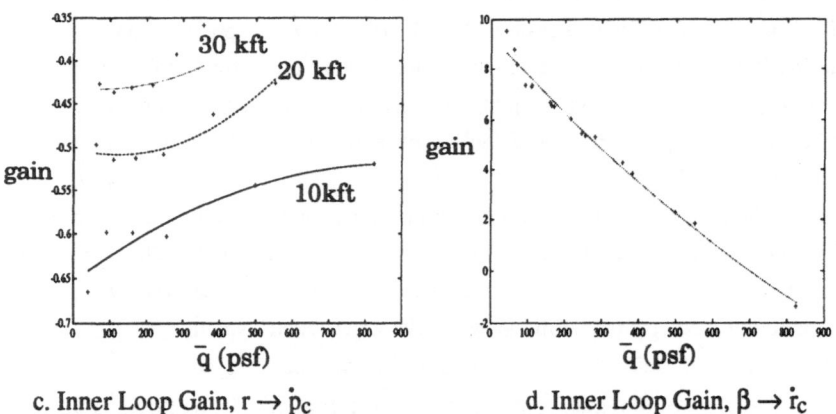

c. Inner Loop Gain, $r \rightarrow \dot{p}_c$ d. Inner Loop Gain, $\beta \rightarrow \dot{r}_c$

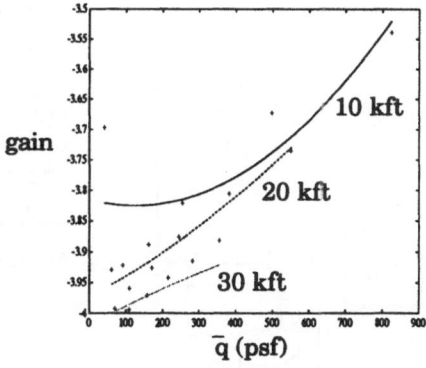

e. Inner Loop Gain, $r \rightarrow \dot{r}_c$

Fig. 5.25 (cont.) Inner Loop Gains

5.4.2 Outer Loop Design

Since the inner loop feedback gains and the control selector are both scheduled with flight condition, the inner loop dynamics are invariant with respect to flight condition. For the outer loop feedback design, a nominal flight condition is chosen, so nominal plant dynamics and the corresponding inner loop feedback gains can be used in the synthesis model. The flight condition at Mach 0.5 and 20,000 feet altitude is chosen as the nominal design point. The dynamic pressure at this condition is 170 psf, and the trim angle of attack is 5.4 degrees. The control effectiveness matrix is replaced by the generalized control effectiveness given in eq. (5.5).

The synthesis model used for the outer loop control law design is shown in Fig. 5.26.

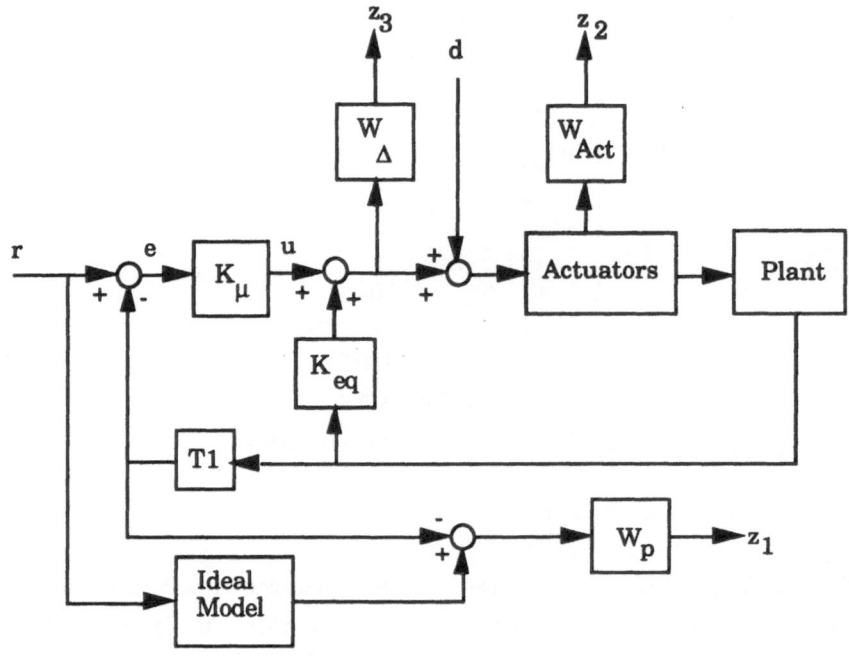

Fig. 5.26 Outer Loop Design Model

Included in the model are the nominal plant model including the control selector, the inner loop feedback gains, K_{eq}, and the output transformation, T1, given by

$$T1 = \begin{bmatrix} 1 & 0 & 0 \\ 0 & \cos\alpha & \sin\alpha \end{bmatrix}. \tag{5.32}$$

Performance is achieved by including an ideal model of the roll rate and sideslip angle responses in the synthesis model. By using target flying qualities parameters, the ideal model is:

$$\frac{\beta_{ideal}}{\beta_c} = \frac{9}{(s^2 + 4.2s + 9)}, \qquad \frac{\dot{\mu}_{ideal}}{\dot{\mu}_c} = \frac{1}{(0.6s + 1)}. \tag{5.33}$$

To achieve performance, an error signal is generated between the ideal model and the actual model responses. This error is weighted with frequency dependent weights, shown in Fig. 5.27, and given by

$$W_p = diag(W_p^{\beta}, W_p^{\mu}),$$

$$W_p^{\beta} = \frac{0.05(s + 500)}{(s + 0.1)} \qquad W_p^{\mu} = \frac{0.005(s + 500)}{(s + 0.1)}. \qquad (5.34)$$

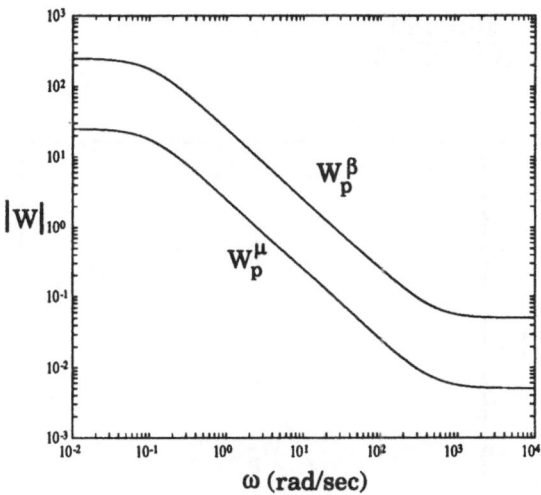

Fig. 5.27 Performance Weights

The input matrix is replaced by the generalized input matrix in the nominal design model to represent the control selector matrix. Because of this, the real control inputs are replaced by generalized control inputs of commanded roll and yaw acceleration. Since the real actuator models cannot be included in this arrangement, the actuator models used in the design are fictitious, and are given by

$$\frac{\delta}{\delta_c} = \frac{20}{(s + 20)}. \qquad (5.35)$$

To prevent unrealistic control commands, the actuator rates are weighted in the problem formulation by

$$W_{act} = 0.005 \, I_2. \qquad (5.36)$$

Robustness to unstructured uncertainty is achieved by including a multiplicative uncertainty at the plant input. The uncertainty bounds are shown in Fig. 5.28, and are given by

$$W_\Delta = \text{diag}(W_{\dot{p}}, W_{\dot{r}}),$$

$$W_{\dot{p}} = W_{\dot{r}} = \frac{10(s + 5)}{(s + 50,000)}.$$

(5.37)

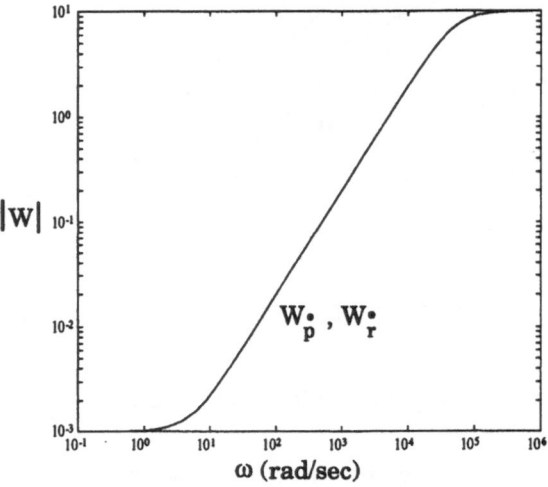

Fig. 5.28 Uncertainty Bounds

The synthesis model is rearranged to form the interconnection structure in Fig. 5.29. Here, z_1 is the weighted error between the ideal model and the output of the plant, z_2 is the weighted actuator rate vector, and z_3 is the output from the unstructured uncertainty scaling W_Δ. The perturbation block includes a 2 by 4 performance block between r and z_1 and z_2 and a 2 by 2 uncertainty block between d and z_3.

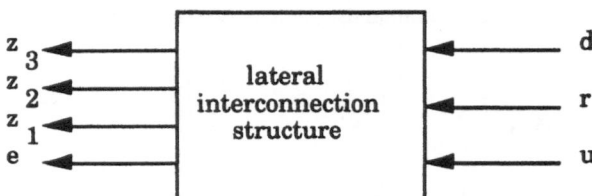

Fig. 5.29 Lateral Interconnection Structure

μ-synthesis is applied to the interconnection structure by first designing an H_∞ control law. The initial bound on the H_∞ norm is 1.325. μ-analysis is performed on the closed loop system, and third order D-scaling matrices are fit to the frequency dependent D scales used to find the upper bound of μ. These are used to scale the original plant. An H_∞ control law was found for the scaled plant, with an H_∞ bound of 0.552. μ-analysis was performed again, and third order D-scales were used a second time. The final control law was found with an H_∞ norm bound of 0.545.

The resulting μ-synthesis controller has 24 state variables and is stable. The controller is balanced and reduced to 10 state variables by truncating the 14 state variables with the smallest Hankel singular values. The resulting 10th order controller includes 2 large negative poles, one of which is -2399.5, the other of which is -472.8. They are residualized to form the final 8th order controller. The full-order and reduced-order controllers are given in Appendix 5.

5.4.3 Robustness Analysis

Although the control laws are designed with a specific accounting for unstructured uncertainty at the plant input to account for the uncertainty in the actuators, there are many sources of uncertainty in the dynamic aircraft model that were not accounted for directly in the design. To try to account directly for all of the sources of uncertainty would be computationally intensive and result in a very large order controller. Robustness must be verified against all sources of uncertainty after the design is completed to assess the need for redesign with more rigorous treatment of different types of uncertainty.

While fictitious control inputs, actuator models, and uncertainty descriptions are used in the design process to be able to incorporate the control selector into the nominal plant model, the actual control input matrix, dynamic actuator models, and uncertainty models of the actuators are used in the analysis of robustness. The weights used for the unstructured uncertainty at the plant input are plotted in Fig. 5.30, and are:

$$W_R = W_{DT} = \frac{s}{(s + 30)}, \qquad W_A = \frac{(s + 0.1)}{(s + 100)}. \qquad (5.38)$$

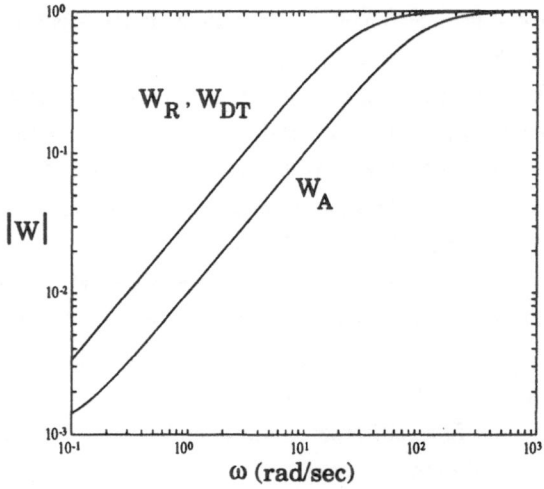

Fig. 5.30 Weights for Actuator Uncertainty

Note that only the aerodynamic actuators are used in the analysis for the same reasons discussed in Section 5.3.3.

In addition to unstructured uncertainty at the input to reflect the modeling errors and neglected dynamics in the actuators, there is also corresponding unstructured uncertainty at the output of the plant to reflect the modeling errors and neglected dynamics in the sensors. The measurements assumed for this design are the body axis roll and yaw rate signals, and the sideslip angle signal. Although it is now practical to generate a sideslip angle signal for feedback control, the signal is often corrupted with noise and often has significant errors. The robustness of the closed loop system to errors at the plant output is also tested. The weights for the unstructured uncertainty at the plant output are plotted in Fig. 5.31, and are:

$$W_p = W_r = \frac{s}{(s + 200)}, \qquad W_\beta = \frac{10(s + 10)}{(s + 1000)}. \qquad (5.39)$$

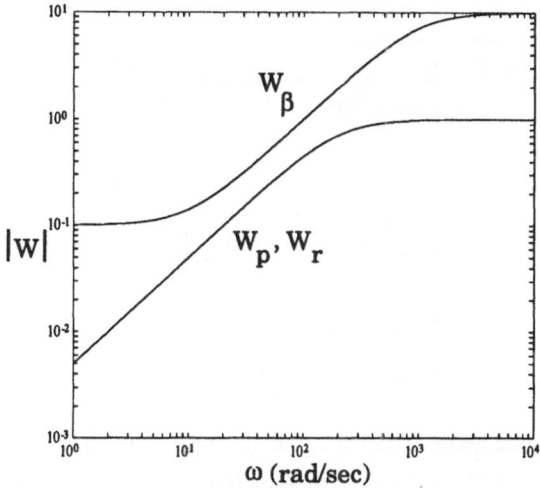

Fig. 5.31 Weights for Sensor Uncertainty

Finally, the stability and control derivatives in the linear state space model are uncertain to various degrees. Errors in modeling and wind tunnel testing, as well as variations in moments of inertia and mass with fuel burn all contribute to the uncertainty in the stability and control derivatives. The level of uncertainty assumed for each stability or control derivative is shown in Table 5.1, and is based on confidence in each of the derivatives. Each uncertainty is given as a percentage of the nominal value.

The interconnection structure for the robustness analysis consists of the eighth order outer loop regulator, the inner loop feedback gain matrix, the control selector matrix, the actuator models, including a fourth order linear model to represent the elevator and second order linear models to represent the ailerons and the rudder, and the third order lateral/directional model. The third order rather than the fourth order lateral/directional model is used because of the marginally stable spiral mode. If the structured singular value were evaluated with the

Table 5.1 Lateral/Directional Parameter Uncertainty Bounds

Stability Derivatives		Control Derivatives	
Derivative	% Uncertainty	Derivative	% Uncertainty
Y_β	15	$Y_{\delta R}$	15
L_β	10	$L_{\delta DT}$	15
L_p	30	$L_{\delta A}$	10
L_r	20	$L_{\delta R}$	40
N_β	30	$N_{\delta DT}$	15
N_p	50	$N_{\delta A}$	20
N_r	15	$N_{\delta R}$	15

marginally stable spiral mode included in the dynamics, it would be very large, since very small perturbations would make this mode unstable. This would corrupt the analysis since an unstable spiral mode can be acceptable with respect to Level 1 flying qualities.

The structured singular value is assessed at four different flight conditions to determine robust stability. The flight conditions used for robustness analysis appear in Appendix 5. The perturbation matrix included a complex 3 by 3 matrix for the unstructured uncertainty at the input, a complex 3 by 3 matrix for the unstructured uncertainty at the plant output, and 14 scalar uncertainties for the uncertain stability and control derivatives. First, the scalar perturbations to the stability and control derivatives are assumed to be complex. Since the parameter values are always real in the actual plant, this analysis is conservative. Next, the scalar perturbations are constrained to be real. This analysis is much less conservative.

The structured singular values are plotted in Fig. 5.32 for each of the four analysis conditions. Where the parameter perturbations are assumed to be complex, the plots are shown as solid lines, and for parameter perturbations assumed to be real, the plots are shown as dashed lines.

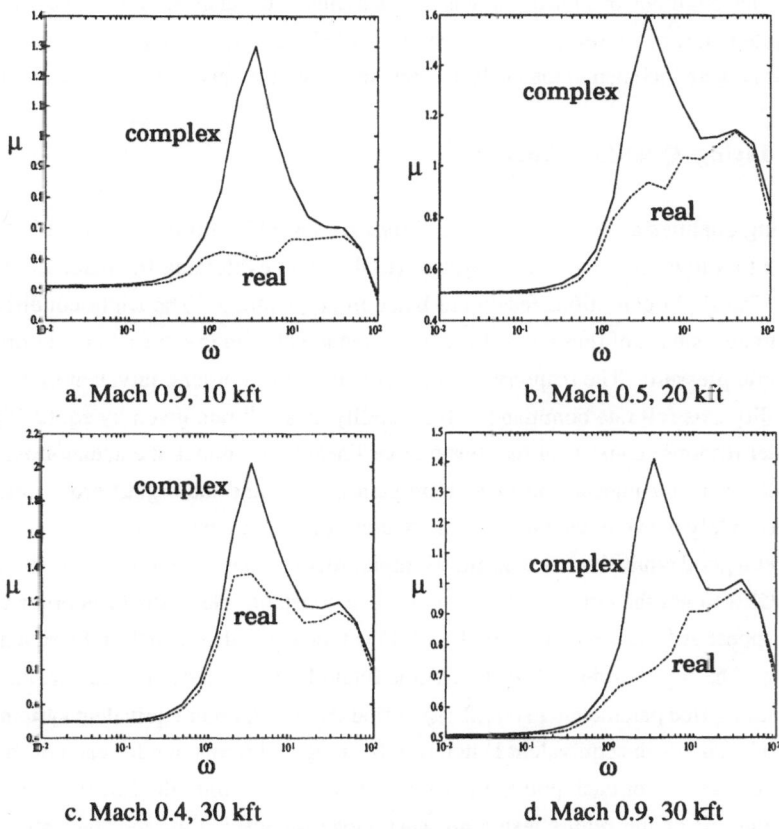

a. Mach 0.9, 10 kft   b. Mach 0.5, 20 kft

c. Mach 0.4, 30 kft   d. Mach 0.9, 30 kft

Fig. 5.32 Robust Stability Plots

At each flight condition, the structured singular value shows a large peak near 3 rad/sec when the parameter uncertainty is assumed to be complex. This peak corresponds to some perturbation that will make the closed loop Dutch roll poles near that frequency unstable. The peaks range from 1.3 to 2.0, meaning that the true plant can only tolerate between 1/1.3 and 1/2.0 of the allowed perturbation.

When the structured singular value is computed with the parameter perturbations constrained to be real, no peaks appeared in the structured singular value near 3 rad/sec, meaning that the parameter perturbation that the plant is sensitive to is complex, a situation which can never physically occur. The maximum values of the structured singular values now lie between 0.65 and 1.3. For two of the cases shown, the value of μ does exceed unity, meaning that for some combination of parameter perturbations, as well as perturbations in the actuators and sensors, the plant will go unstable. Although robustness for all of the simultaneous uncertainty is not guaranteed, the analysis shows that the closed loop system is not extremely sensitive to any of the uncertainty, as many types of uncertainty were included in the analysis, and the value of μ only exceeded one slightly.

5.4.4   Flying Qualities Analysis

The flying qualities at several flight conditions are tested by fitting the actual high order response to a low order equivalent system (LOES) and evaluating the parameters of the LOES. The flight conditions tested are listed in Appendix 5. The flight conditions are chosen to be a subset of those used for design, and are chosen to represent the entire range of dynamic pressure. The response of interest is the fourth order equivalent system from the stability axis roll rate command to the stability axis roll rate given by eq. (2.19). The high order response consists of the fourth order linear plant model, the actuator dynamics, the control selector matrix, the inner loop gain matrix, and the eighth order outer loop controller. Only the aerodynamic control effectors are used in the analysis.

To get a good equivalent system fit, the high order roll rate response is first fit to a first order LOES to get the equivalent roll mode time constant. Next, the high order sideslip angle response is fit to a second order LOES to get the equivalent Dutch roll frequency and damping. The fourth order roll rate LOES is found by fixing these parameters and using the remaining free parameters in eq. (2.19) to find the fourth order equivalent system.

Fig. 5.33 shows the equivalent Dutch roll damping and frequency for each of the flight conditions tested. For each point, the parameters are well within the Level 1 boundaries. In addition, all of the points tested are very close together, although the points tested represent a wide range of dynamic pressure. This is because of the gain scheduled inner loop. Figs. 5.34 and 5.35 show the equivalent roll mode time constant and equivalent time delay, respectively, each plotted as a function of dynamic pressure.

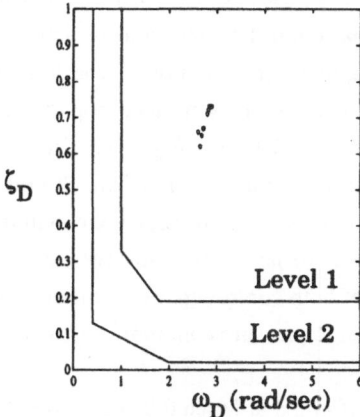

Fig. 5.33 Dutch Roll Frequency and Damping

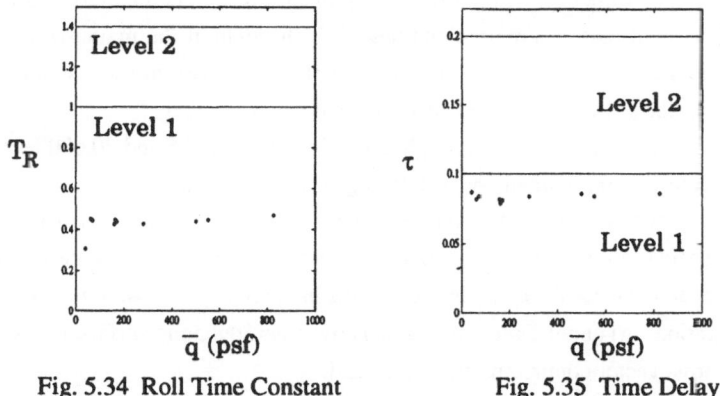

Fig. 5.34 Roll Time Constant   Fig. 5.35 Time Delay

Similar to the Dutch roll parameters, each of these is well within the Level 1 boundaries, and each is fairly constant with dynamic pressure because of the gain scheduled inner loop.

5.5 Nonlinear Analysis

Using the nonlinear simulation with the control laws implemented as FORTRAN subroutines, several maneuvers are simulated. First the outer loop controller blending, described in section 5.3.2, is analyzed with angle of attack unit step commands given to the longitudinal control system. Next, three highly coupled maneuvers are performed and analyzed to evaluate the control system during realistic operating conditions.

Angle of attack step inputs are used to demonstrate outer loop controller blending performance of the longitudinal control system during nonlinear simulations at high, intermediate, and low dynamic pressure conditions. The three different conditions are chosen such that three different combinations of outer loop controller commands are used based on the structure shown in Fig. 5.16 and Fig. 5.17. Fig. 5.36 shows the nonlinear responses of the aircraft model to identical angle of attack unit step inputs at the three different conditions. The high and low dynamic pressure conditions use only the high and low dynamic pressure controllers, respectively. However, the middle dynamic pressure condition uses a combination of both controllers resulting in an angle of attack response between the high and low dynamic pressure conditions. Note that thrust vectoring is not required for this benign maneuver.

Next, a 3g loaded roll is performed at Mach 0.8 and 20,000 feet. This maneuver is a coupled maneuver that excites dynamic modes in the lateral/directional and longitudinal axes. The maneuver consists of the aircraft entering a 3g turn, with a roll reversal to a 3g turn in the opposite direction without unloading the aircraft. The maneuver is shown in Fig. 5.37. The load factor is maintained near 3g's throughout the maneuver, while the sideslip angle never exceeds 1.5 degrees. This maneuver does not saturate the aerodynamic surfaces, therefore thrust vectoring is not commanded.

A high rate roll is performed at the flight condition of Mach 0.5 and 20,000 feet. A 250 deg/sec roll rate is commanded for a 360 deg. roll. The state responses and control deflections are plotted in Fig. 5.38. The stability axis roll rate response is satisfactory with good turn coordination in that sideslip remains less than 2 deg. Thrust vector commands are quickly generated since the ailerons and rudders immediately rate saturate to the point of position saturation. At about 3 sec into the maneuver, aerodynamic surface rate saturation ceases, and thrust vector commands are not needed.

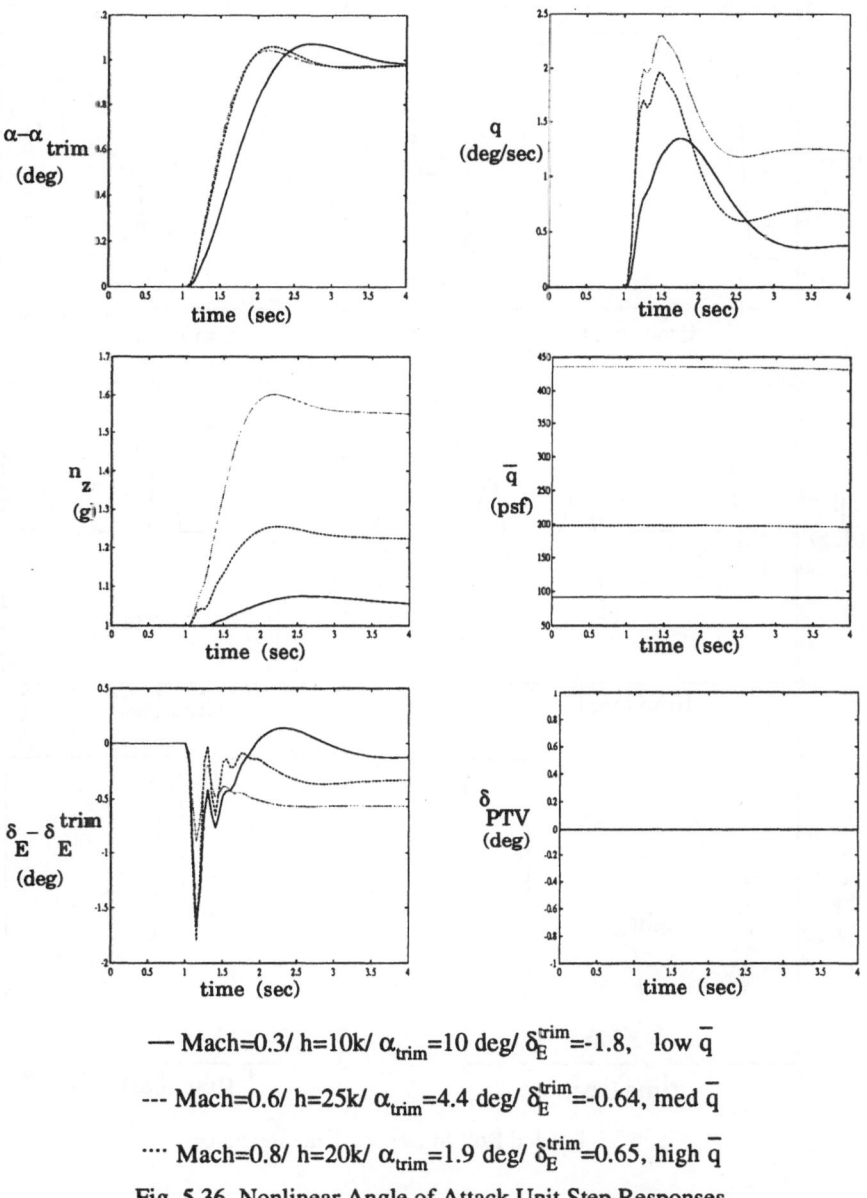

— Mach=0.3/ h=10k/ α_{trim}=10 deg/ δ_E^{trim}=-1.8, low $\bar{q}$

--- Mach=0.6/ h=25k/ α_{trim}=4.4 deg/ δ_E^{trim}=-0.64, med $\bar{q}$

···· Mach=0.8/ h=20k/ α_{trim}=1.9 deg/ δ_E^{trim}=0.65, high $\bar{q}$

Fig. 5.36 Nonlinear Angle of Attack Unit Step Responses

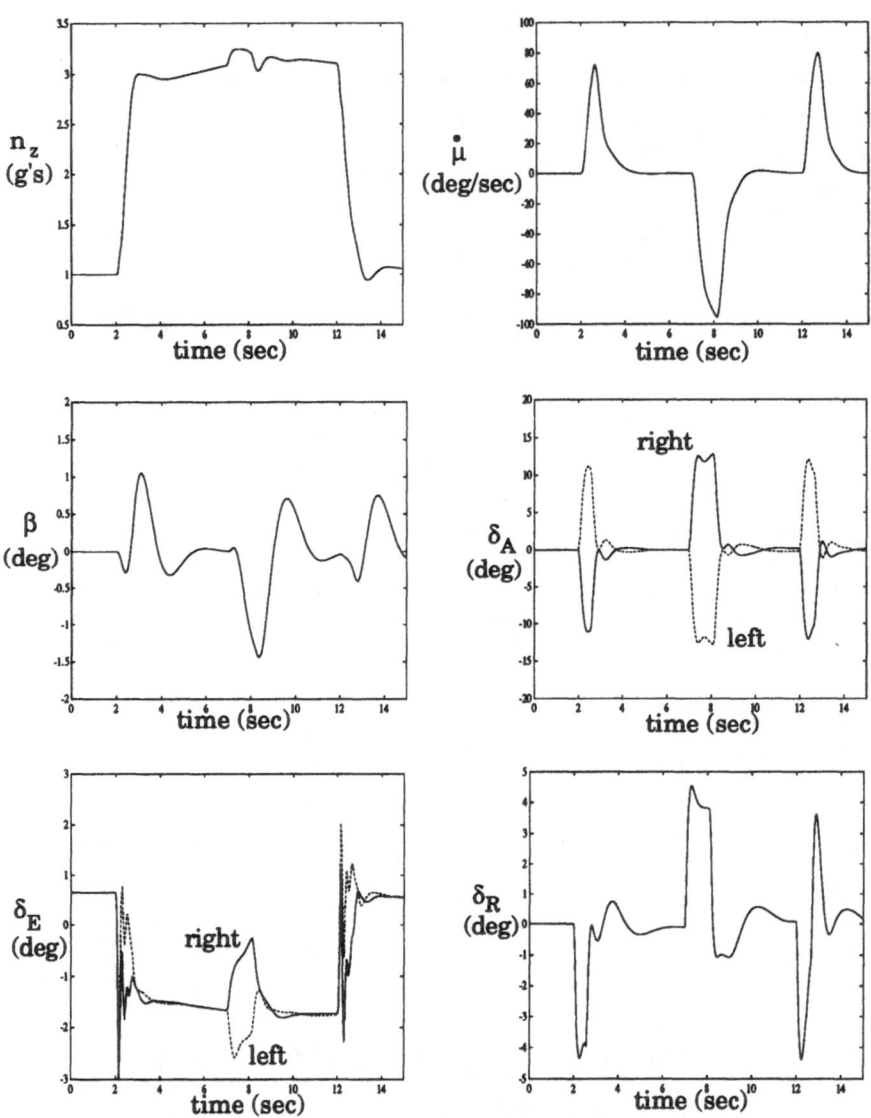

Fig. 5.37 Loaded Roll Maneuver Time Response

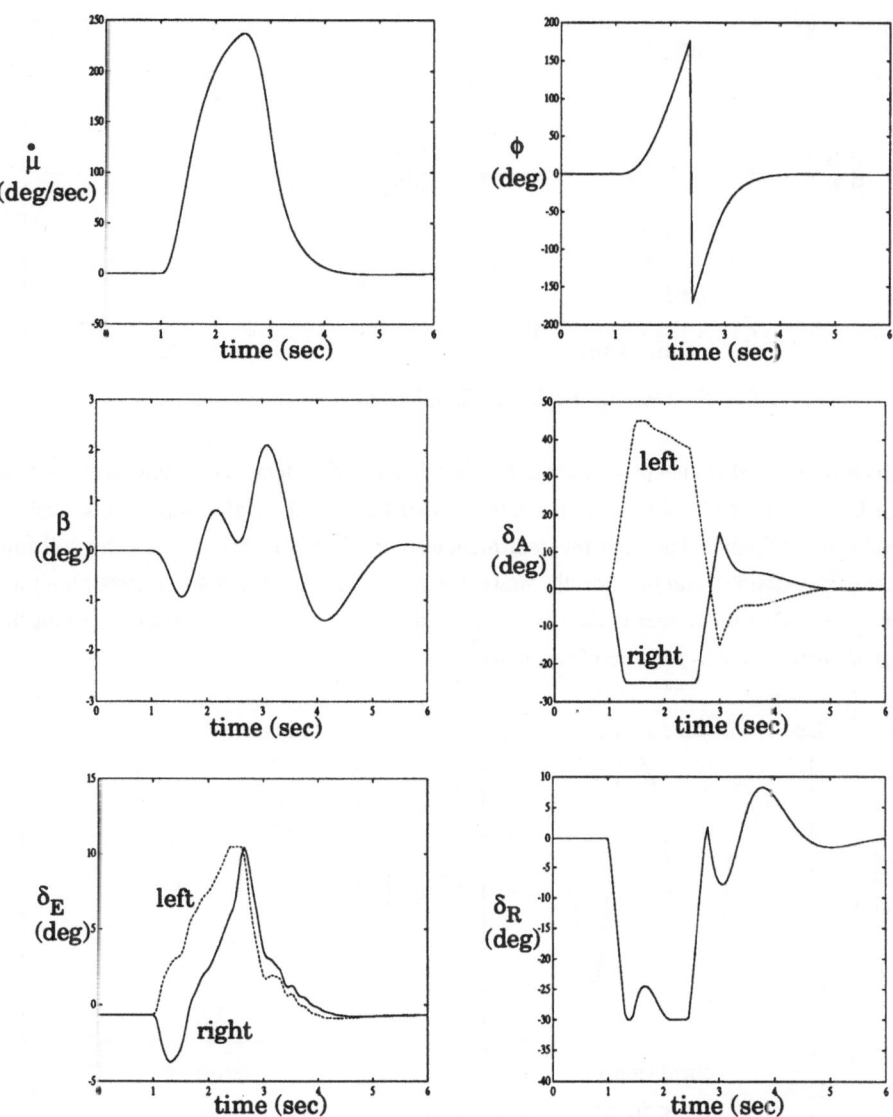

Fig. 5.38 High Rate Roll Maneuver Time Response

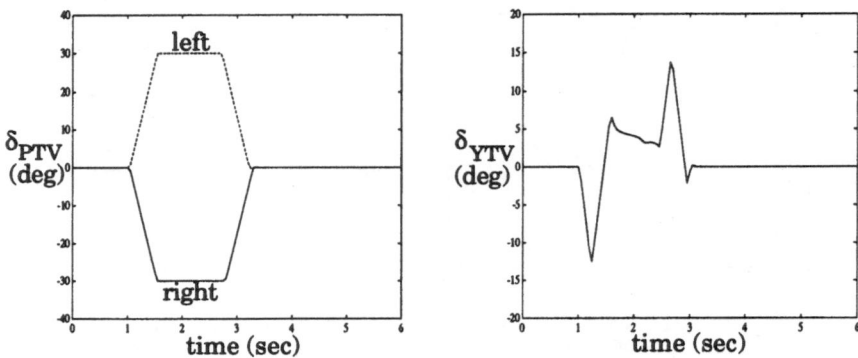

Fig. 5.38 (cont.) High Rate Roll Maneuver Time Response

A turn reversal is also performed at Mach 0.6 and 20,000 feet. This maneuver consists of a bank to the right, a bank to the left, another bank to the right, and a bank back to steady, level flight. The turn reversal maneuver is plotted in Fig. 5.39. The roll rate responds as desired, and the sideslip angle is maintained at less than 4 deg throughout the maneuver. This maneuver is also extreme enough to warrant thrust vectoring about the lateral/directional axes because of aerodynamic surface saturation.

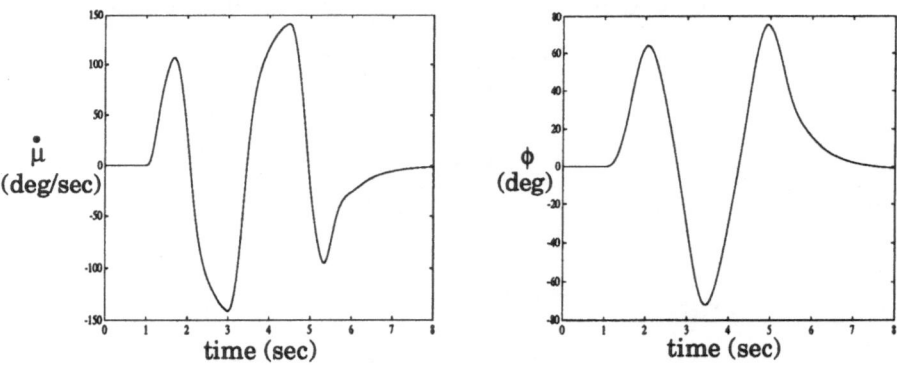

Fig. 5.39 Turn Reversal Maneuver Time Response

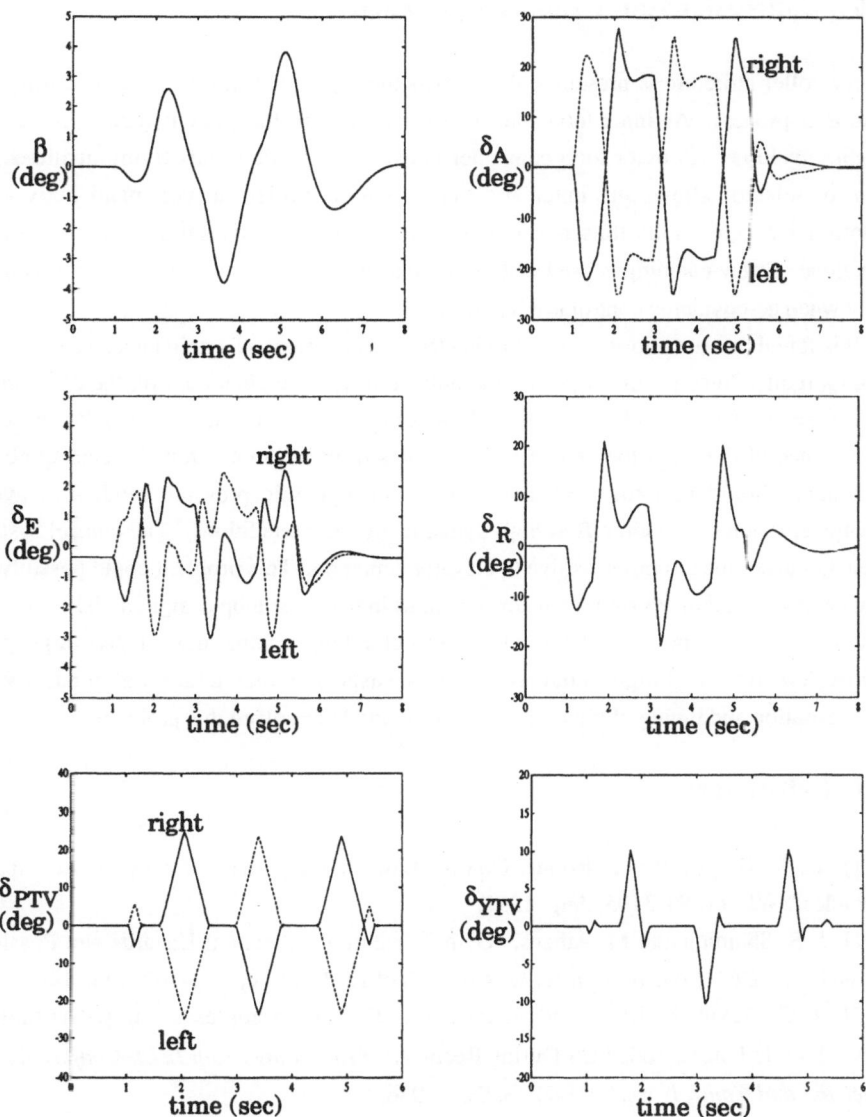

Fig. 5.39 (cont.) Turn Reversal Maneuver Time Response

Note that stability and performance of the aircraft are maintained even though the high rate roll and turn reversal maneuvers are extreme enough to saturate the thrust vectoring nozzles.

5.6 Conclusions and Lessons Learned

A controller structure is presented that allows the separation of design goals during the synthesis process. An inner loop controller is used to equalize plant dynamics across the flight envelope. An outer loop controller is used to provide robust flying qualities. A control selector allows the inner and outer loop controllers to command body axes rotational accelerations thus normalizing the control effectiveness across the flight envelope. Daisy-chaining is used within the control selector to select thrust vector control only when aerodynamic control is not sufficient.

It is found that the prioritization of redundant control effectors within the control selector is important. Since the elevators are the only aerodynamic pitch control, the differential elevator commands must be reduced to allow for symmetric elevator commands if needed. The choice of the inner loop equalized dynamics turns out to be crucial. The equalized dynamics should be chosen similar to the high dynamic pressure models to avoid deaugmentation of the aircraft at high dynamic pressure conditions. The control system seems to make the aircraft sensitive to elevator actuation. This problem could possibly be reduced by including elevator actuator dynamics in the outer loop design model. It is also noted that the methods used for inner and outer loop control design give high gain controllers for the longitudinal axis. It is suspected that additional feedforward compensation and tuning of the design weights could have solved this problem.

5.7 References

[5.1] K. R. Haiges et al., "Robust Control Law Development for Modern Aerospace Vehicles," WL-TR-91-3105, Aug. 1991.

[5.2] J. S. Shamma and M. Athans, "Gain Scheduling: Potential Hazards and Possible Remedies," *IEEE Control Systems Magazine*, Vol.12, No.3, pp.101-107, June 1992.

[5.3] J. C. Doyle, K. Lenz, and A. Packard, "Design Examples Using μ-Synthesis: Space Shuttle Lateral Axis FCS During Reentry," *Proceedings 25th IEEE Conference on Decision and Control*, pp. 2218-2223, Dec. 1986.

Appendix 5

This section gives the model data of the design presented in the previous sections. First, detailed actuator models are given, then linear models and information about the flight conditions at which the linear models are generated are given. Inner loop control gain schedules are given, and finally, full and reduced order outer loop dynamical controllers are presented.

ACTUATOR MODELS

High-order detailed aerodynamic actuator models are reduced to simple 4th-order and 2nd-order models, shown in Table A5.1, to avoid very small time-step integration during nonlinear simulation.

Table A5.1 Actuator Models

Surface	Actuator Dynamics	Rate Limits (deg/sec)	Position Limits (deg)
Elevator	$((\frac{s}{82.9})^2 + \frac{2(0.068)}{82.9}s + 1.0, [((\frac{s}{36.4})^2 + \frac{2(0.41)}{36.4}s + 1.0] [(\frac{s}{105.3})^2 + \frac{2(0.59)}{105.3}s + 1.0])$	-60.0 +60.0	-24.0 +10.5
Leading Edge Flap	$\dfrac{1.0}{[\frac{s}{26.9} + 1.0] [\frac{s}{82.9} + 1.0]}$	-40.0 +40.0	0.0 +35.0
Trailing Edge Flap	$\dfrac{1.0}{(\frac{s}{35.0})^2 + \frac{2(0.71)}{35.0}s + 1.0}$	-100.0 +100.0	-10.0 +45.0
Aileron	$\dfrac{1.0}{(\frac{s}{75.0})^2 + \frac{2(0.59)}{75.0}s + 1.0}$	-100.0 +100.0	-25.0 +45.0
Rudder	$\dfrac{1.0}{(\frac{s}{72.0})^2 + \frac{2(0.69)}{72.0}s + 1.0}$	-100.0 +100.0	-30.0 +30.0
Vectoring Nozzles	$\dfrac{1.0}{(\frac{s}{20.0})^2 + \frac{2(0.60)}{20.0}s + 1.0}$	-60.0 +60.0	-30.0 +30.0

Thrust-vectoring nozzle actuators are assumed to be 2nd-order filters and are also given in Table A5.1. These reduced-order models are used in the nonlinear simulation and linear robustness analysis.

LINEAR DESIGN MODELS AND FLIGHT CONDITIONS

The flight conditions that are chosen for linear control design, flying qualities analysis, and robustness analysis are given in Table A5.2. The conditions chosen represent a broad range of dynamic pressures throughout the envelope since inner-loop gains are scheduled versus dynamic pressure.

Table A5.2 Flight Conditions
a. Longitudinal Design Conditions

Mach Number	Altitude (ft)	$\bar{q}$ (psf)	α (deg)	Flying Qualities Analysis	Robustness Analysis
0.3	26000	47.4	25.2	√	
0.5	40000	68.5	16.8	√	√
0.6	30000	158.4	5.2	√	
0.4	6000	189.9	6.0	√	√
0.7	14000	426.4	2.6	√	√
0.8	12000	603.0	1.9	√	√
0.95	20000	614.4	1.6	√	
0.8	10000	652.0	1.7	√	
0.8	5000	789.1	1.5	√	
0.9	10000	825.2	1.4	√	
0.85	5000	890.8	1.4	√	
0.9	5000	998.7	1.3	√	√

Table A5.2 (cont.)

b. Lateral/Directional Design Conditions

Mach Number	Altitude (ft)	$\bar{q}$ (psf)	α (deg)	Flying Qualities Analysis	Robustness Analysis
0.2	10,000	40.75	29.7	√	
0.3	10,000	91.69	10.0		
0.4	10,000	163.00	5.6	√	
0.5	10,000	254.68	3.6		
0.7	10,000	499.18	1.8	√	
0.9	10,000	825.17	1.2	√	√
0.3	20,000	61.27	12.6	√	
0.4	20,000	108.92	7.8		
0.5	20,000	170.19	5.4	√	√
0.6	20,000	245.07	3.9		
0.75	20,000	382.92	2.8		
0.9	20,000	551.40	2.1	√	
0.4	30,000	70.39	17.4	√	√
0.5	30,000	109.98	8.1		
0.6	30,000	158.37	5.2	√	
0.7	30,000	215.55	3.6		
0.8	30,000	281.54	2.2	√	
0.9	30,000	356.32	1.5		√

The linear models for each flight condition of Table A5.2 are listed below, where the nomenclature, for example A_{long}^{m9h10}, is the longitudinal state matrix at Mach .9 and 10 kft.

The longitudinal linear models have the following form:

$$\begin{bmatrix} \dot{\alpha} \\ \dot{q} \end{bmatrix} = \begin{bmatrix} Z_\alpha & Z_q \\ M_\alpha & M_q \end{bmatrix} \begin{bmatrix} \alpha \\ q \end{bmatrix} + \begin{bmatrix} Z_{\delta E} & Z_{\delta PTV} \\ M_{\delta E} & M_{\delta PTV} \end{bmatrix} \begin{bmatrix} \delta_E \\ \delta_{PTV} \end{bmatrix}$$

$$= A_{long} \begin{bmatrix} \alpha \\ q \end{bmatrix} + B_{long} \begin{bmatrix} \delta_E \\ \delta_{PTV} \end{bmatrix} \qquad\qquad (A5.1)$$

$$A_{long}^{m3h26} = \begin{bmatrix} -0.2296 & 0.9931 \\ 0.02436 & -0.2046 \end{bmatrix} \qquad B_{long}^{m3h26} = \begin{bmatrix} -0.04034 & -0.01145 \\ -1.73 & -0.517 \end{bmatrix}$$

$$A_{\text{long}}^{m5h40} = \begin{bmatrix} -0.2423 & 0.9964 \\ -2.342 & -0.1737 \end{bmatrix} \qquad B_{\text{long}}^{m5h40} = \begin{bmatrix} -0.0416 & -0.01141 \\ -2.595 & -0.8161 \end{bmatrix}$$

$$A_{\text{long}}^{m6h30} = \begin{bmatrix} -0.5088 & 0.994 \\ -1.131 & -0.2804 \end{bmatrix} \qquad B_{\text{long}}^{m6h30} = \begin{bmatrix} -0.09277 & -0.01787 \\ -6.573 & -1.525 \end{bmatrix}$$

$$A_{\text{long}}^{m4h6} = \begin{bmatrix} -0.8018 & 0.9847 \\ -1.521 & -0.5944 \end{bmatrix} \qquad B_{\text{long}}^{m4h6} = \begin{bmatrix} -0.1508 & -0.02776 \\ -7.926 & -1.751 \end{bmatrix}$$

$$A_{\text{long}}^{m7h14} = \begin{bmatrix} -1.175 & 0.9871 \\ -8.458 & -0.8776 \end{bmatrix} \qquad B_{\text{long}}^{m7h14} = \begin{bmatrix} -0.194 & -0.03593 \\ -19.29 & -3.803 \end{bmatrix}$$

$$A_{\text{long}}^{m8h12} = \begin{bmatrix} -1.562 & 0.9862 \\ -14.94 & -1.132 \end{bmatrix} \qquad B_{\text{long}}^{m8h12} = \begin{bmatrix} -0.2316 & -0.04349 \\ -26.48 & -5.323 \end{bmatrix}$$

$$A_{\text{long}}^{m95h20} = \begin{bmatrix} -1.905 & 0.9895 \\ -33.88 & -0.9872 \end{bmatrix} \qquad B_{\text{long}}^{m95h20} = \begin{bmatrix} -0.1867 & -0.03287 \\ -27.22 & -4.573 \end{bmatrix}$$

$$A_{\text{long}}^{m8h10} = \begin{bmatrix} -1.675 & 0.9853 \\ -16.16 & -1.212 \end{bmatrix} \qquad B_{\text{long}}^{m8h10} = \begin{bmatrix} -0.2449 & -0.04649 \\ -28.34 & -5.742 \end{bmatrix}$$

$$A_{\text{long}}^{m8h5} = \begin{bmatrix} -1.994 & 0.9828 \\ -19.44 & -1.427 \end{bmatrix} \qquad B_{\text{long}}^{m8h5} = \begin{bmatrix} -0.2852 & -0.05567 \\ -33.44 & -6.931 \end{bmatrix}$$

$$A_{\text{long}}^{m9h10} = \begin{bmatrix} -2.452 & 0.9856 \\ -38.61 & -1.34 \end{bmatrix} \qquad B_{\text{long}}^{m9h10} = \begin{bmatrix} -0.2757 & -0.05226 \\ -37.36 & -7.247 \end{bmatrix}$$

$$A_{\text{long}}^{m85h5} = \begin{bmatrix} -2.328 & 0.9831 \\ -30.44 & -1.493 \end{bmatrix} \qquad B_{\text{long}}^{m85h5} = \begin{bmatrix} -0.3012 & -0.05866 \\ -38.43 & -7.815 \end{bmatrix}$$

$$A_{\text{long}}^{m9h5} = \begin{bmatrix} -2.911 & 0.9835 \\ -46.47 & -1.553 \end{bmatrix} \qquad B_{\text{long}}^{m9h5} = \begin{bmatrix} -0.3161 & -0.06231 \\ -43.65 & -8.752 \end{bmatrix}$$

The lateral/directional linear models have the following form:

$$\begin{bmatrix} \dot{\beta} \\ \dot{p} \\ \dot{r} \end{bmatrix} = \begin{bmatrix} Y_\beta & \sin\alpha & -\cos\alpha \\ L_\beta & L_p & L_r \\ N_\beta & N_p & N_r \end{bmatrix} \begin{bmatrix} \beta \\ p \\ r \end{bmatrix} +$$

$$\begin{bmatrix} Y_{\delta DT} & Y_{\delta A} & Y_{\delta R} & Y_{\delta RTV} & Y_{\delta YTV} \\ L_{\delta DT} & L_{\delta A} & L_{\delta R} & L_{\delta RTV} & L_{\delta YTV} \\ N_{\delta DT} & N_{\delta A} & N_{\delta R} & N_{\delta RTV} & N_{\delta YTV} \end{bmatrix} \begin{bmatrix} \delta_{DT} \\ \delta_A \\ \delta_R \\ \delta_{RTV} \\ \delta_{YTV} \end{bmatrix}$$

$$= A_{\text{lat/dir}} \begin{bmatrix} \beta \\ p \\ r \end{bmatrix} + B_{\text{lat/dir}} \begin{bmatrix} \delta_{DT} \\ \delta_A \\ \delta_R \\ \delta_{RTV} \\ \delta_{YT} \end{bmatrix} \qquad\qquad (A5.2)$$

$$A_{\text{lat/dir}}^{m2h10} = \begin{bmatrix} -0.05904 & 0.4959 & -0.8703 \\ -5.513 & -0.9391 & 0.6655 \\ 0.06838 & 0.02632 & -0.1038 \end{bmatrix}$$

$$B_{\text{lat/dir}}^{m2h10} = \begin{bmatrix} 0.005629 & 0.005764 & 0.003685 & 0 & 0.0904 \\ 1.879 & 1.328 & 0.02922 & 0.6754 & 0.217 \\ -0.1092 & -0.09645 & -0.08404 & 0.006811 & -2.974 \end{bmatrix}$$

$$A_{lat/dir}^{m3h10} = \begin{bmatrix} -0.1292 & 0.1738 & -0.9833 \\ -8.643 & -1.129 & 0.5986 \\ 1.519 & -0.01327 & -0.1105 \end{bmatrix}$$

$$B_{lat/dir}^{m3h10} = \begin{bmatrix} -0.006987 & -0.005249 & 0.01285 & 0 & 0.006894 \\ 5.096 & 6.075 & 0.51 & 0.1781 & 0.02478 \\ 0.1908 & -0.1522 & -0.3872 & 0.0008849 & -0.3397 \end{bmatrix}$$

$$A_{lat/dir}^{m4h10} = \begin{bmatrix} -0.1544 & 0.09691 & -0.9939 \\ -9.965 & -1.721 & 0.599 \\ 2.169 & -0.01995 & -0.1447 \end{bmatrix}$$

$$B_{lat/dir}^{m4h10} = \begin{bmatrix} -0.01187 & -0.006276 & 0.01785 & 0 & 0.002272 \\ 9.643 & 12.16 & 0.9326 & 0.1495 & 0.01088 \\ 0.2768 & -0.2727 & -0.7155 & -0.000305 & -0.1492 \end{bmatrix}$$

$$A_{lat/dir}^{m5h10} = \begin{bmatrix} -0.1932 & 0.06234 & -0.9968 \\ -12.37 & -2.164 & 0.6034 \\ 3.199 & -0.0211 & -0.1802 \end{bmatrix}$$

$$B_{lat/dir}^{m5h10} = \begin{bmatrix} -0.01652 & -0.007459 & 0.02203 & 0 & 0.002821 \\ 15.18 & 18.04 & 1.412 & 0.3035 & 0.01689 \\ 0.3458 & -0.3975 & -1.099 & -0.002071 & -0.2315 \end{bmatrix}$$

$$A_{lat/dir}^{m7h10} = \begin{bmatrix} -0.2701 & 0.03162 & -0.9984 \\ -17.77 & -3.177 & 0.5446 \\ 5.987 & -0.0205 & -0.2555 \end{bmatrix}$$

$$B_{lat/dir}^{m7h10} = \begin{bmatrix} -0.02472 & -0.00764 & 0.02855 & 0 & 0.01074 \\ 29.4 & 25.95 & 2.503 & 1.789 & 0.09002 \\ 0.4006 & -0.3672 & -1.982 & -0.02069 & -1.234 \end{bmatrix}$$

$$A_{lat/dir}^{m9h10} = \begin{bmatrix} -0.321 & 0.02008 & -0.9987 \\ -17.6 & -5.716 & 0.5193 \\ 9.433 & -0.02149 & -0.3391 \end{bmatrix}$$

$$B_{lat/dir}^{m9h10} = \begin{bmatrix} -0.0301 & 0 & 0.03051 & 0 & 0.02583 \\ 46.68 & 19.41 & 4.054 & 5.024 & 0.2782 \\ 0.2385 & -0.3226 & -2.951 & -0.06746 & -3.813 \end{bmatrix}$$

$$A_{lat/dir}^{m3h20} = \begin{bmatrix} -0.0714 & 0.2991 & -0.9539 \\ -6.746 & -0.5918 & 0.4968 \\ 0.4099 & 0.003356 & -0.08297 \end{bmatrix}$$

$$B_{lat/dir}^{m3h20} = \begin{bmatrix} -0.00274 & -0.002723 & 0.006669 & 0 & 0.03367 \\ 2.784 & 2.627 & 0.2294 & 0.5298 & 0.1165 \\ 0.04708 & -0.1018 & -0.2059 & 0.005966 & -1.597 \end{bmatrix}$$

$$A_{lat/dir}^{m4h20} = \begin{bmatrix} -0.112 & 0.1408 & -0.9889 \\ -8.538 & -1.171 & 0.5146 \\ 1.619 & -0.01304 & -0.103 \end{bmatrix}$$

$$B_{lat/dir}^{m4h20} = \begin{bmatrix} -0.007388 & -0.004613 & 0.01238 & 0 & 0.008751 \\ 6.245 & 7.673 & 0.611 & 0.3615 & 0.04034 \\ 0.1998 & -0.1822 & -0.4722 & 0.0008652 & -0.553 \end{bmatrix}$$

$$A_{lat/dir}^{m5h20} = \begin{bmatrix} -0.1354 & 0.09036 & -0.9949 \\ -10.37 & -1.469 & 0.5126 \\ 2.281 & -0.01482 & -0.1277 \end{bmatrix}$$

$$B_{lat/dir}^{m5h20} = \begin{bmatrix} -0.01091 & -0.005695 & 0.01555 & 0 & 0.00489 \\ 9.93 & 12.12 & 0.9416 & 0.3977 & 0.02817 \\ 0.2757 & -0.2797 & -0.7419 & -0.001175 & -0.3861 \end{bmatrix}$$

$$A_{lat/dir}^{m6h20} = \begin{bmatrix} -0.166 & 0.0629 & -0.9971 \\ -12.97 & -1.761 & 0.5083 \\ 3.191 & -0.01417 & -0.1529 \end{bmatrix}$$

$$B_{lat/dir}^{m6h20} = \begin{bmatrix} -0.0142 & -0.00686 & 0.01851 & 0 & 0.005817 \\ 14.38 & 16.76 & 1.316 & 0.7007 & 0.0402 \\ 0.3389 & -0.385 & -1.051 & -0.00475 & -0.5511 \end{bmatrix}$$

$$A_{\text{lat/dir}}^{m75h20} = \begin{bmatrix} -0.1982 & 0.03905 & -0.9984 \\ -17.3 & -2.505 & 0.4624 \\ 4.688 & -0.01064 & -0.1942 \end{bmatrix}$$

$$B_{\text{lat/dir}}^{m75h20} = \begin{bmatrix} -0.01816 & -0.007716 & 0.02168 & 0 & 0.01018 \\ 22.53 & 23.71 & 1.905 & 1.697 & 0.08794 \\ 0.4211 & -0.238 & -1.555 & -0.01765 & -1.206 \end{bmatrix}$$

$$A_{\text{lat/dir}}^{m9h20} = \begin{bmatrix} -0.2257 & 0.02638 & -0.9989 \\ -19.08 & -3.708 & 0.4264 \\ 6.586 & -0.01925 & -0.2379 \end{bmatrix}$$

$$B_{\text{lat/dir}}^{m9h20} = \begin{bmatrix} -0.02089 & -0.005593 & 0.02314 & 0 & 0.01809 \\ 31.29 & 25.88 & 2.879 & 3.359 & 0.1875 \\ 0.3459 & -0.1457 & -2.152 & -0.04183 & -2.571 \end{bmatrix}$$

$$A_{\text{lat/dir}}^{m4h30} = \begin{bmatrix} -0.07253 & 0.2183 & -0.9751 \\ -6.216 & -0.6013 & 0.427 \\ 0.9358 & -0.005054 & -0.07243 \end{bmatrix}$$

$$B_{\text{lat/dir}}^{m4h30} = \begin{bmatrix} -0.003475 & -0.003369 & 0.007457 & 0 & 0.02451 \\ 3.491 & 3.867 & 0.3446 & 0.606 & 0.1084 \\ 0.09003 & -0.115 & -0.2766 & 0.004285 & -1.486 \end{bmatrix}$$

$$A_{\text{lat/dir}}^{m5h30} = \begin{bmatrix} -0.09336 & 0.1351 & -0.99 \\ -8.39 & -0.9786 & 0.437 \\ 1.629 & -0.009743 & -0.0886 \end{bmatrix}$$

$$B_{\text{lat/dir}}^{m5h30} = \begin{bmatrix} -0.006711 & -0.003948 & 0.01047 & 0 & 0.01242 \\ 6.183 & 7.5 & 0.5958 & 0.6251 & 0.06865 \\ 0.2107 & -0.1819 & -0.475 & 0.001079 & -0.9411 \end{bmatrix}$$

$$A_{\text{lat/dir}}^{m6h30} = \begin{bmatrix} -0.1118 & 0.09363 & -0.9949 \\ -10.22 & -1.169 & 0.4318 \\ 2.201 & -0.009853 & -0.1056 \end{bmatrix}$$

$$B_{\text{lat/dir}}^{m6h30} = \begin{bmatrix} -0.009231 & -0.004885 & 0.01262 & 0 & 0.008328 \\ 9.02 & 11.06 & 0.842 & 0.7304 & 0.05522 \\ 0.2957 & -0.2637 & -0.6841 & -0.001928 & -0.757 \end{bmatrix}$$

$$A_{\text{lat/dir}}^{m7h30} = \begin{bmatrix} -0.1275 & 0.06791 & -0.997 \\ -11.92 & -1.404 & 0.4283 \\ 2.776 & -0.01099 & -0.124 \end{bmatrix}$$

$$B_{\text{lat/dir}}^{m7h30} = \begin{bmatrix} -0.01108 & -0.005726 & 0.01428 & 0 & 0.007537 \\ 12.54 & 14.81 & 1.085 & 0.9813 & 0.0583 \\ 0.3656 & -0.2088 & -0.9165 & -0.005925 & -0.7992 \end{bmatrix}$$

$$A_{\text{lat/dir}}^{m8h30} = \begin{bmatrix} -0.1397 & 0.04954 & -0.9981 \\ -14.78 & -1.859 & 0.3924 \\ 3.437 & -0.003867 & -0.1429 \end{bmatrix}$$

$$B_{\text{lat/dir}}^{m8h30} = \begin{bmatrix} -0.01269 & -0.006594 & 0.01585 & 0 & 0.009622 \\ 16.56 & 18.24 & 1.372 & 1.527 & 0.08506 \\ 0.3773 & -0.08791 & -1.168 & -0.01347 & -1.166 \end{bmatrix}$$

$$A_{\text{lat/dir}}^{m9h30} = \begin{bmatrix} -0.1542 & 0.03692 & -0.9987 \\ -16.19 & -2.384 & 0.3588 \\ 4.42 & -0.01486 & -0.1624 \end{bmatrix}$$

$$B_{\text{lat/dir}}^{m9h30} = \begin{bmatrix} -0.01401 & -0.006381 & 0.01653 & 0 & 0.01256 \\ 20.34 & 20.8 & 1.879 & 2.187 & 0.1249 \\ 0.3122 & -0.03341 & -1.472 & -0.02367 & -1.712 \end{bmatrix}$$

INNER LOOP GAIN SCHEDULES

The longitudinal inner loop equalization gain schedules are given by:

$$F = -40 \qquad K_f = [1 \quad 1] \qquad T = [0 \quad .0247 \quad 0]$$

$$N = N(\bar{q}) \qquad M = [M_1(\bar{q}) \quad M_2(\bar{q})].$$

$$N(\bar{q}) = -.312\,\bar{q} + 461$$

$$M_1(\bar{q}) = -.058\,\bar{q} + 50.5 \qquad M_2(\bar{q}) = -.006\,\bar{q} + 8.11 \qquad \textbf{(A5.3)}$$

where $\bar{q}$ is the dynamic pressure. The lateral/directional inner loop equalization gain schedules are given by

$$\text{gain} = c_2\,\bar{q}^2 + c_1\,\bar{q} + c_0 , \qquad\qquad \textbf{(A5.4)}$$

where the coefficients, c_i, are given in Table A5.3.

OUTER LOOP CONTROLLERS

The full-order and reduced-order outer loop robust performance controllers have the following form:

$$K_\mu = \frac{\left[\begin{array}{cc} \text{num11} & \text{num12} \\ \text{num21} & \text{num22} \end{array}\right]}{\text{den}},$$

$$\text{num} = K(s + z_1)(s + z_2)\cdots(s + z_{n-1})(s + z_n)$$

$$\text{den} = (s + p_1)(s + p_2)\cdots(s + p_{n-1})(s + p_n) . \qquad \textbf{(A5.5)}$$

The poles, zeros, and high frequency gains of eq. (A5.5) are given in Tables A5.4, A5.5, A5.6, and A5.7.

Table A5.3 Inner Loop Lateral/Directional Gain Functions

Gain	Altitude	c_2	c_1	c_0
$\beta \rightarrow \dot{p}_c$	all	-3.7683e-05	4.8235e-02	3.5702e+00
$p \rightarrow \dot{p}_c$	all	2.2130e-06	4.3849e-03	-1.1015e+00
$r \rightarrow \dot{p}_c$	10 kft	-1.7065e-07	3.0285e-04	-6.5368e-01
$r \rightarrow \dot{p}_c$	20 kft	4.7299e-07	-1.1692e-04	-5.0102e-01
$r \rightarrow \dot{p}_c$	30 kft	3.4239e-07	-5.3058e-05	-4.3078e-01
$\beta \rightarrow \dot{r}_c$	all	4.3901e-06	-1.6425e-02	9.3445e+00
$r \rightarrow \dot{r}_c$	10 kft	6.2272e-07	-1.5514e-04	-3.8156e+00
$r \rightarrow \dot{r}_c$	20 kft	2.3004e-07	3.1432e-04	-3.9730e+00
$r \rightarrow \dot{r}_c$	30 kft	-1.3339e-07	3.2521e-04	-4.0197e+00

Table A5.4 Full Order Outer Loop Longitudinal Controller Terms

Terms	High $\bar{q}$ Controller	Low $\bar{q}$ Controller
K	3.2460e+01	1.1662e+01
z_1	-1.0000e+04	-1.0000e+04
z_2	-5.6966e+01	-7.8015e+01
z_3	-3.9957e+01	-3.9997e+01
z_4	-3.2275e+01	-9.1515e+00+ j8.1670e+00
z_5	-1.1415e+01+ j9.8971e+00	-9.1515e+00- j8.1670e+00
z_6	-1.1415e+01- j9.8971e+00	-4.3756e+00+ j7.3457e+00
z_7	-9.1025e+00+ j7.2716e+00	-4.3756e+00- j7.3457e+00
z_8	-9.1025e+00- j7.2716e+00	-5.5259e+00+ j5.2578e+00
z_9	-5.5028e+00+ j5.3695e+00	-5.5259e+00- j5.2578e+00
z_{10}	-5.5028e+00- j5.3695e+00	-5.5596e-01
z_{11}	-8.9949e-01	-8.9794e+00
z_{12}	-1.9290e+00	-1.2393e+00
p_1	-1.0090e+02	-1.0050e+02
p_2	-5.6976e+01	-7.8015e+01
p_3	-5.4479e+01	-7.7265e+01
p_4	-3.8398e+01	-3.9991e+01
p_5	-3.2131e+01	-1.3962e+01+ j1.0358e+01
p_6	-2.8024e+01	-1.3962e+01- j1.0358e+01
p_7	-2.1437e+01	-4.2595e+00+ j7.1293e+00
p_8	-8.5408e+00+ j8.0166e+00	-4.2595e+00- j7.1293e+00
p_9	-8.5408e+00- j8.0166e+00	-8.6239e+00
p_{10}	-8.2418e+00	-3.6453e-02
p_{11}	-7.2782e-02	-1.2349e+00
p_{12}	-8.8269e-01	-4.7961e+00
p_{13}	-1.9109e+00	-5.4663e-01

Table A5.5 Reduced-Order Outer Loop Longitudinal Controller Terms

Coefficients	High $\bar{q}$ Controller	Low $\bar{q}$ Controller
K	1.1995e+02	1.2951e+02
z_1	-2.4180e+03	-6.0168e+02
z_2	-5.4039e+00+ j7.2572e+00	-3.8407e+00+ j6.2507e+00
z_3	-5.4039e+00- j7.2572e+00	-3.8407e+00- 6.2507e+00j
p_1	-7.1836e+01+ j4.0476e+01	-5.9135e-01- j5.1963e+01
p_2	-7.1836e+01- j4.0476e+01	-5.9135e+01+ j5.1963e+01
p_3	-7.2319e-02	-3.6321e-02
p_4	-2.0509e+01	-6.6426e+00

Table A5.6 Full-Order Outer Loop Lateral/Directional Controller Terms

Terms	num11	num12	num21	num22	den
K (p_1)	1.5684e+01	4.9687e-02	-1.2148e+02	2.3936e-01	-2.4172e+03
z_1 (p_2)	-5.0000e+04	-5.0000e+04	-5.0000e+04	-5.0000e+04	-4.6906e+02
z_2 (p_3)	-4.6906e+02	-3.8500e+03	-4.6906e+02	-4.6905e+02	-4.6906e+02
z_3 (p_4)	-8.1594e+01	-4.6906e+02	-8.0915e+01	-1.1422e+02	-8.1594e+01
z_4 (p_5)	-8.0915e+01	-8.0915e+01	-8.1594e+01	-8.0915e+01	-8.1594e+01
z_5 (p_6)	-4.7416e+01+j1.9234e+01	-8.1594e+01	4.7331e+01+j1.9169e+01	-8.1594e+01	4.7271e+01+j1.8037e+01
z_6 (p_7)	-4.7416e+01-j1.9234e+01	-4.7271e+01+j1.8037e+01	-4.7331e+01-j1.9169e+01	-4.7421e+01+j2.0238e+01	-4.7271e+01-j1.8037e+01
z_7 (p_8)	4.7271e+01+j1.8037e+01	-4.7271e+01-j1.8037e+01	4.7271e+01+j1.8037e+01	-4.7421e+01-j2.0238e+01	4.7271e+01+j1.8037e+01
z_8 (p_9)	-4.7271e+01-j1.8037e+01	4.7849e+01+j2.0195e+01	-4.7271e+01-j1.8037e+01	4.7271e+01+j1.8037e+01	-4.7271e+01-j1.8037e+01
z_9 (p_{10})	-1.8854e+01	-4.7849e+01-j2.0195e+01	-1.5469e+01	-4.7271e+01-j1.8037e+01	5.4460e+00+j7.0157e+00
z_{10} (p_{11})	-1.1416e+01	-1.9301e+01	5.5529e+00+j7.1005e+00	-1.4946e+01	-5.4460e+00-j7.0157e+00
z_{11} (p_{12})	4.2560e+00+j5.7923e+00	5.0577e+00+j6.5979e+00	-5.5529e+00-j7.1005e+00	3.1802e+00+j7.0493e+00	5.1303e+00+j6.6357e+00
z_{12} (p_{13})	-4.2560e+00-j5.7923e+00	-5.0577e+00-j6.5979e+00	4.7010e+00+j3.5892e+00	-3.1802e+00-j7.0493e+00	-5.1303e+00-j6.6357e+00
z_{13} (p_{14})	4.8355e+00+j3.6809e+00	5.0731e+00+j1.0852e+00	-4.7010e+00-j3.5892e+00	4.4362e+00+j2.0233e+00	-6.8492e+00
z_{14} (p_{15})	-4.8355e+00-j3.6809e+00	-5.0731e+00-j1.0852e+00	-6.8363e+00	-4.4362e+00-j2.0233e+00	5.0910e+00+j1.0723e+00
z_{15} (p_{16})	-5.7669e+00	-4.8788e+00	2.3675e+00+j2.5835e+00	-5.9915e+00	-5.0910e+00-j1.0723e+00
z_{16} (p_{17})	-3.1217e+00	-3.1217e+00	-2.3675e+00-j2.5835e+00	-3.8357e+00	-3.1167e+00
z_{17} (p_{18})	-3.1247e+00	-3.1160e+00	-3.1222e+00	-3.1298e+00	-3.1220e+00
z_{18} (p_{19})	3.3081e-01	-1.7925e+00	-3.1217e+00	-3.1217e+00	-4.5012e-03
z_{19} (p_{20})	-6.0353e-04	-1.8030e-02	-7.6581e-03	-6.6312e-03	-5.7684e-02
z_{20} (p_{21})	-1.8030e-02	-1.8279e-02	-1.8064e-02	-5.7550e-02	-7.6719e-03
z_{21} (p_{22})	-1.7979e-02	-4.4855e-03	-5.7726e-02	-1.7970e-02	-5.7726e-02
z_{22} (p_{23})	-5.7737e-02	-5.7682e-02	-5.7726e-02	-5.7726e-02	-1.8030e-02
z_{23} (p_{24})	-5.7726e-02	-5.7726e-02	-1.8030e-02	-1.8030e-02	-1.8030e-02

Table A5.7 Reduced-Order Outer Loop Lateral/Directional Controller Terms

Terms	num11	num12	num21	num22	den
K (-)	6.6247e-01	-1.0744e-03	-5.1999e+00	8.4320e-03	-
z_1 (p_1)	-1.5580e+01+j4.9885e+00	7.4783e+03	-1.6466e+01	-1.7658e+02	-5.3683e+00+j5.8269e+00
z_2 (p_2)	-1.5580e+01-j4.9885e+00	-1.8158e+01	5.2682e+00+j7.8519e+00	3.9919e+00+j8.8339e+00	-5.3683e+00-j5.8269e+00
z_3 (p_3)	3.4145e+00+j6.1475e+00	5.4242e+00+j5.8610e+00	-5.2682e+00-j7.8519e+00	-3.9919e+00-j8.8339e+00	5.2608e+00+j7.8938e+00
z_4 (p_4)	-3.4145e+00-j6.1475e+00	-5.4242e+00-j5.8610e+00	2.2273e+00+j3.0615e+00	4.3893e+00+j2.1244e+00	-5.2608e+00-j7.8938e+00
z_5 (p_5)	3.3817e-01	-1.6278e+00	-2.2273e+00-j3.0615e+00	-4.3893e+00-j2.1244e+00	-4.4999e-03
z_6 (p_6)	-5.4958e-02	-5.2459e-02	-5.2431e-02	-5.3326e-02	-7.6457e-03
z_7 (p_7)	-6.0643e-04	-2.5573e-02	-2.5330e-02	-2.4672e-02	-2.5310e-02
z_8 (p_8)	-2.4413e-02	-4.4841e-03	-7.6325e-03	-6.6453e-03	-5.2293e-02

CHAPTER 6

CONCLUSIONS

This document provides theory and examples for a design methodology for manual flight control system design. The inner/outer loop structure gives a versatile framework for the application of advanced multivariable control theory to aircraft control problems. The engineer has the freedom to choose any method which achieves the design goals. This freedom is demonstrated in the successful application of three different inner loop equalization approaches in the VISTA F-16 and supermaneuverable vehicle design examples. This framework is also versatile enough to include a broad range of design specific issues such as the management of redundant control effectors, flying qualities requirements, modeling uncertainties, and nonlinearities.

While the design examples presented in this document illustrate a broad range of design issues, they still do not represent flight ready control laws. Issues not addressed in this study that are worthy of further attention include: turbulence responses, sensor noise, changing inertial properties due to different loadings and fuel burn, redundancy, and the details of pilot interfaces.

INDEX